BIOMIMICRY: NATURE-INSPIRED INNOVATIONS

BIOMIMICRY

YAGESH KUMAR AND PRABIR SARKAR

This book is didicated

To almighty god, who have been beside us the whole time, and always gave us helpful advice to complete our work for their unwavering support and encouragement throughout writing this book.

For the natural world, intricate creations just make you astonished and brainstorming morely in different feild as it is an area that provides inspiration to the engineering field.

In this respect, this book shows an integration of natural make-up of a knowledgeable person and the intelligence of a human.

Contents

Abstract

Abstract

Biomimicry, an emerging interdisciplinary approach, seeks to emulate and adapt nature's time-tested strategies to address complex engineering and technological challenges. This book explores the integration of biomimetic principles into engineering design processes. Inspired by the remarkable adaptations found in biological systems, this study delves into the structural, functional, and material aspects of various organisms, including plants, animals, and microorganisms. Leveraging insights from nature's optimization, the book endeavors to develop innovative solutions for sustainability, efficiency, and resilience in engineering applications. Through a meticulous analysis of biomimetic case studies and principles, this research aims to showcase the potential of biomimicry as a transformative tool for enhancing sustainability and efficiency across diverse engineering domains.

This book encompasses a comprehensive literature review, identifying key biomimetic principles and success stories across multiple fields such as materials sciences, and sustainable design. The core focus is on developing biomimetic design methodologies that integrate biological insights into the engineering design process

Case studies include the development of lightweight structural materials inspired by bone architecture, modeled after animal behaviors, and energy-efficient building designs inspired by the cooling mechanisms of termite mounds. The book emphasizes sustainability, efficiency, and ecological harmony as key outcomes, aligning with the global shift toward environmentally responsible engineering practices.

The results demonstrate that biomimicry can foster groundbreaking advancements in engineering by leveraging nature's timeless innovations. By fostering interdisciplinary collaboration between biologists, engineers, and designers, this book highlights the potential of biomimicry as a transformative force in engineering and innovation, paving the way for a more sustainable and harmonious future.

Biomimicry, an interdisciplinary approach, seeks to emulate nature's principles and solutions to address complex engineering challenges. This book explores the integration of biomimicry in engineering, aiming to revolutionize sustainable design practices. The book delves into bio-inspired design methodologies, studying biological systems, structures, and processes to derive innovative engineering solutions. Key focus areas include material science, aerodynamics, and robotics, where insights from nature guide the development of efficient and sustainable technologies. Through rigorous research and experimentation, this study demonstrates the potential of biomimicry to drive advancements in engineering, contributing to a more harmonious relationship between technology and the natural world.

Foreword

Welcome to the world of bio-inspired design, where engineering meets nature, and the possibilities are as boundless as the natural world itself.

Engineering has always been a field of human endeavor that combines the elegance of natural principles with the need of scientific methodology. As in today world we can say that the boundaries between biology and engineering are increasingly blurring which means the boundary is decreasing day by day, giving rise to innovative solutions inspired by the natural world. This branch or field is often referred to as bio-inspired or biomimetic engineering, advances the efficiency, adaptability, and resilience found in biological systems to address complex engineering challenges.

In this book, we explore the fascinating intersection of biology and engineering, uncovering how nature's time-tested designs can inspire groundbreaking advancements in technology and industry. From the intricacies of a bird's wing informing aerodynamics to the strength and flexibility of spider silk influencing new materials, bio-inspired design offers a wealth of insights that can revolutionize traditional engineering approaches.

The principles of bio-inspired design are lying in the observation and understanding of natural systems. By studying how organisms have evolved to solve problems related to structure, movement, energy efficiency, and environmental adaptation, engineers can develop innovative solutions that are not only effective but also sustainable. For example, the microscopic structure of lotus leaves has inspired self-cleaning surfaces, and the unique properties of shark skin have led to the development of drag-reducing materials for aquatic applications.

This book is divided into several sections, each focusing on a specific area where bio-inspired design has made significant impacts. We begin with an overview of the fundamental concepts of biomimicry and the methods used to translate biological phenomena into engineering solutions. Subsequent chapters delve into case studies from various fields, including aerospace, robotics, materials science, and environmental engineering, showcasing real-world applications and the tangible benefits they provide.

Through this exploration, we aim to inspire a new generation of engineers and scientists to look beyond traditional boundaries and embrace the creativity and ingenuity that nature offers. By doing so, we can pave the way for a future where human innovation and natural inspiration go hand in hand, leading to a more sustainable and resilient world.

Welcome to the world of bio-inspired design, where engineering meets nature, and the possibilities are as boundless as the natural world itself.

Preface

Biomimicry, an emerging interdisciplinary approach, seeks to emulate and adapt nature's time-tested strategies to address complex engineering and technological challenges. This M.Tech book explores the integration of biomimetic principles into engineering design processes. Inspired by the remarkable adaptations found in biological systems, this study delves into the structural, functional, and material aspects of various organisms, including plants, animals, and microorganisms. Leveraging insights from nature's optimization, the book endeavors to develop innovative solutions for sustainability, efficiency, and resilience in engineering applications. Through a meticulous analysis of biomimetic case studies and principles, this research aims to showcase the potential of biomimicry as a transformative tool for enhancing sustainability and efficiency across diverse engineering domains.

This book encompasses a comprehensive literature review, identifying key biomimetic principles and success stories across multiple fields such as materials sciences, and sustainable design. The core focus is on developing biomimetic design methodologies that integrate biological insights into the engineering design process

Case studies include the development of lightweight structural materials inspired by bone architecture, modeled after animal behaviors, and energy-efficient building designs inspired by the cooling mechanisms of termite mounds. The book emphasizes sustainability, efficiency, and ecological harmony as key outcomes, aligning with the global shift toward environmentally responsible engineering practices.

The results demonstrate that biomimicry can foster groundbreaking advancements in engineering by leveraging nature's timeless innovations. By fostering interdisciplinary collaboration between biologists, engineers, and designers, this book highlights the potential of biomimicry as a transformative force in engineering and innovation, paving the way for a more sustainable and harmonious future.

Biomimicry, an interdisciplinary approach, seeks to emulate nature's principles and solutions to address complex engineering challenges. This book explores the integration of biomimicry in engineering, aiming to revolutionize sustainable design practices. The book delves into bio-inspired design methodologies, studying biological systems, structures, and processes to derive innovative engineering solutions. Key focus areas include material science, aerodynamics, and robotics, where insights from nature guide the development of efficient and sustainable technologies. Through rigorous research and experimentation, this study demonstrates the potential of biomimicry to drive advancements in engineering, contributing to a more harmonious relationship between technology and the natural world.

From the authors

This book is mainly the outcome of extensive research carried out by Yagesh, a Masters student of the department of mechanical engineering at IIT Ropar. The works was carried out under the supervision of Dr. Prabir, a faculty at IIT Ropar.

We have tried to bring out the basic and latest knowledge in this book on biomimicry. After reading the book, the reader would have fundamental understanding of the amazing works on biomimicry. Biomimicry is also known by many terms such as bio-mimitices, biomimicry, bio inspired etc.

We have tried to add references wherever possible. We used Zotero (https://www.zotero.org/) as the main reference manager tool. In some references, there is no date in the reference, thus the Zotero system has automatically assigned n.d. (that is no date) at the end of some references. We apologise for this error.

Most of the images are taken from internet which are free/ having creative common licence or are royalty free.

We would be happy to receive any feedback on this book. Please feel free to share your feedback to the first author Mr. Yagesh (yageshkumar108@gmail.com)

Acknowledgements

We would like to express our gratitude to the field of biomimicry for inspiring and shaping this work. Biomimicry, also known as nature-inspired design, provided us with a valuable framework for developing innovative solutions by emulating and learning from nature's design principles and strategies.

This is mainly the outcome of the research work carried out by the first author as a M.Tech research. Throughout he explored various biological systems, organisms, and natural processes to address the challenges and complexities that we encountered. By studying the adaptability and efficiency of natural systems, we were able to draw inspiration and apply the principles of biomimicry to our technological solutions. The application of biomimicry in our book has enabled us to design sustainable and eco-friendly solutions that are aligned with the natural world. We have learned how nature functions as a sustainable ecosystem, and by incorporating these principles into our book, we hope to contribute towards a more sustainable and harmonious future.

We would like to extend our gratitude to the researchers, scientists, and pioneers in the field of biomimicry who have paved the way for us to explore and integrate nature's wisdom into our book. Their dedication to studying and understanding the intricate workings of nature has been useful in our book's development.

Finally, we would like to express our appreciation to mentors, advisors, and colleagues who have supported and guided me throughout our book. Their expertise and insights have been invaluable in realizing the application of biomimicry in our book and pushing the boundaries of technological innovation. We hope to continue exploring and utilizing biomimicry principles in our future endeavors to create sustainable and effective solutions for the benefit of both humans and the environment.

This work has been funded by IIT ROPAR, AWADH (Agriculture Water Hub), DST funded and Department of Mechanical Engineering, IIT ROPAR.

Prologue

Biomimicry is more than just a scientific discipline or a technological; it is a profound shift in perspective—a recognition that the natural world is not just a resource to be exploited, but a source of wisdom, innovation, and profound beauty. It is a recognition that, in the elegant forms and intricate processes of living organisms, lie answers to some of the most pressing challenges facing our species.

As we delve into the wonders of biomimicry, we will encounter creatures with extraordinary abilities: from the lotus leaf, whose microscopic structure repels water with unparalleled efficiency, to the gecko, whose feet adhere to surfaces with a strength that defies gravity. We will marvel at the resilience of organisms that have weathered the harshest conditions, from the depths of the desert to the heart of the Arctic. And we will uncover the hidden connections that bind all living things together in a web of life that is as fragile as it is resilient.

But biomimicry is not just about mimicking the forms of nature; it is about understanding the principles that underlie them—the principles of efficiency, adaptability, and sustainability. It is about learning from the wisdom of billions of years of evolution and applying that wisdom to the challenges of our own making.

In the pages ahead, we will explore the ways in which biomimicry is already transforming fields as diverse as architecture, engineering, materials science, and medicine. We will meet the visionaries and innovators who are harnessing the power of nature to revolutionize technology and design. And we will confront the urgent need to embrace a more sustainable and regenerative approach to the way we live, work, and interact with the natural world.

Above all, we will be reminded of the extraordinary potential that lies within each and every one of us—to learn from nature, to be inspired by its beauty and complexity, and to forge a new relationship with the living world that sustains us all. For in the end, the story of biomimicry is not just a story of innovation; it is a story of hope—a reminder that, even in the darkest of times, nature offers us a guiding light, a blueprint for a brighter future.

CHAPTER I

Introduction to biomimicry

This chapter expresses the multifaceted essence of bio-inspiration as an innovative design approach rooted in nature, aiming to address human challenges sustainably across diverse fields. Beginning with a historical overview, it traces the origins of bio-inspiration from ancient civilizations to the modern era, emphasizing its evolution into a comprehensive interdisciplinary practice. The abstract underscores the significance of bio-inspiration in engineering, highlighting its role in deriving efficient and sustainable solutions by replicating nature's intricate designs honed over millions of years. It elaborates on how bio-inspired engineering contributes to sustainability efforts, reduces resource consumption, and fosters eco-friendly practices, aligning with global environmental priorities. Furthermore, the abstract explores the diverse applications of bio-inspiration, from materials science to robotics and medicine, showcasing its potential to revolutionize various industries. Emphasizing the collaborative nature of bio-inspired design, it underscores the importance of interdisciplinary collaboration in fostering cross-disciplinary innovation. Ultimately, the abstract positions bio-inspiration as a beacon of hope in addressing escalating environmental challenges, offering promising pathways towards a more resilient and sustainable future for society.

1.1 Defining Bio-Inspiration

Bio-inspiration, also known as biomimicry, is a design and innovation approach that draws inspiration from nature to solve human challenges and create more sustainable technologies and solutions. It involves studying biological systems, processes, and structures found in the natural world and applying the insights gained to design and engineering to create products, systems, or processes that are more efficient, sustainable, and well-suited to their intended purpose. Bio-inspiration can encompass a wide range of disciplines, from architecture and materials science to robotics and medicine, and it often leads to innovative and environmentally friendly solutions (Bensaude-Vincent, n.d.-a, n.d.-b; Taieb & Amor, n.d.).

1

1.2 Historical Context

The concept of bio-inspired design has its roots in the observation and imitation of nature, dating back to ancient civilizations. However, the modern systematic exploration of bio-inspiration can be traced to the mid-20th century. Here's a brief historical context:

Throughout history, humans have consistently found inspiration in the natural world, utilizing its intricate designs for a range of inventions. This includes endeavors like emulating bird flight to advance aviation and employing natural materials for crafting tools and shelters. The inborn curiosity and keen observation of nature's mechanisms have been enduring aspects of human exploration and creativity. (Bensaude-Vincent, n.d.-a; Kennedy et al., 2015a)

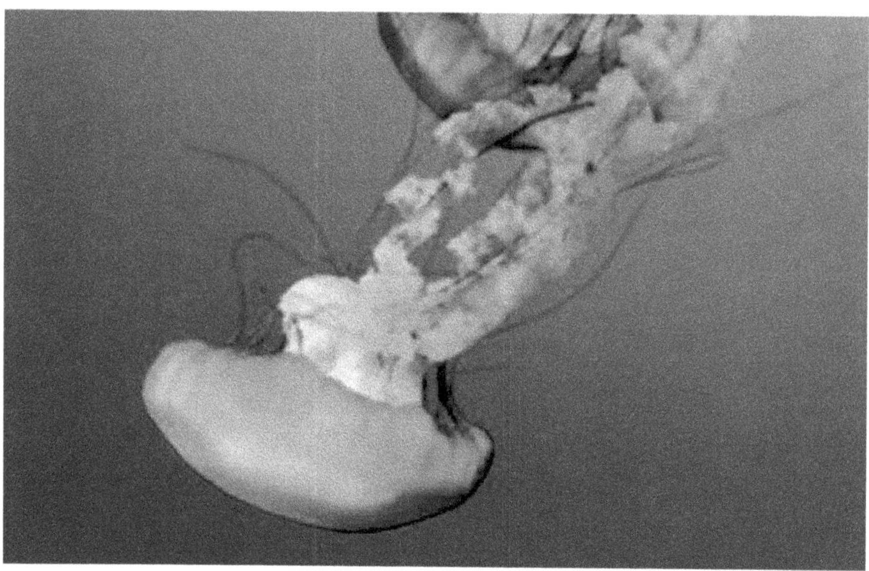

Fig.1.1.1: Trapping into Nature BiomimicryPathak, S. (2019). *Biomimicry: (Innovation Inspired by Nature).*

In the 1960s, NASA took a significant step in formalizing bio-inspiration through the initiation of the Biomimetics program. This initiative aimed to explore how natural processes and organisms could inform and enhance aerospace technology, marking a pivotal moment in the integration of biology into engineering and design practices.(Iouguina et al., 2014; Kennedy et al., 2015a, 2015b)

The late 20th century witnessed the rise of biomimicry as a distinct discipline, championed by various scholars and researchers. Janine Benyus played a crucial role in popularizing this concept through her 1997 book, "Biomimicry: Innovation Inspired by Nature." This period marked the formal acknowledgment of the value of bio-inspiration in fostering innovative solutions across diverse fields.(Iouguina et al., 2014; Kennedy et al., 2015b; Lebdioui, 2022)

Biomimetic applications expanded across disciplines, gaining momentum in architecture, materials science, robotics, and medicine. Examples include the design of energy-efficient buildings modeled after termite mounds and the development of robots inspired by the locomotion of animals.(Taieb & Amor, n.d.)

In the 21st century, with an increasing emphasis on sustainability and eco-friendly technologies, bio-inspired design gained prominence as a means to address environmental challenges. Today, bio-inspired design stands as an interdisciplinary field, wielding a growing influence on various industries. It encourages innovation that not only prioritizes efficiency but also aligns with the principles of sustainability and resilience, all while respecting the delicate balance of the natural world. (Iouguina et al., 2014; Kennedy et al., 2015b)

1.3 Importance of Bio-Inspiration in Engineering

Bio-inspiration plays a crucial role in engineering for several reasons:

Bio-inspired engineering derives numerous benefits from the efficiency and optimization inherent in natural systems, honed over millions of years of evolution. By replicating these intricate designs, engineers can create products and systems that not only outperform traditional counterparts but also exhibit a remarkable reduction in resource consumption, resulting in tangible energy and cost savings. (Kennedy et al., 2015c; Lebdioui, 2022; Pathak, 2019)

Moreover, bio-inspired engineering significantly contributes to sustainability efforts. Drawing inspiration from nature's strategies allows engineers to develop technologies with minimal environmental impact, fostering eco-friendly solutions that reduce waste and conserve precious resources. This emphasis on sustainability aligns with the growing global focus on environmentally conscious practices (Iouguina et al., 2014; Kennedy et al., 2015c). Nature serves as a vast repository of solutions to complex problems, having navigated numerous challenges through evolution. Engineers, by studying biological systems, gain invaluable insights that inspire innovative solutions across various engineering domains, from materials science to aerodynamics, providing a treasure trove of strategies for complex problem-solving (Kennedy et al., 2015c). The adaptability and resilience observed in natural systems also become key attributes incorporated into engineering designs. Products and systems infused with these features can better withstand changing conditions and unforeseen challenges, enhancing their overall performance and longevity (Pathak, 2019).

Fig.1.2: Elf ShelterWithington, D. (2009). *"Elf Shelter" Rainwater Collector*

In the context of medical engineering, bio-inspiration becomes particularly crucial. Designing medical devices and implants that mimic the properties of biological tissues enhances biocompatibility, reducing the likelihood of rejection or adverse reactions and advancing the field of healthcare (Bensaude-Vincent, n.d.-a; Kennedy et al., 2015c; Lebdioui, 2022).

Additionally, the exploration of biomaterials and their properties has led to groundbreaking advancements in materials science. Engineers have developed innovative materials with unique characteristics, including self-healing properties, lightweight composites, and superhydrophobic surfaces, opening new frontiers in material design (Iouguina et al., 2014).

Bio-inspired robotics, influenced by animal locomotion and sensory systems, has given rise to a new generation of robots. These robots exhibit enhanced agility, adaptability, and efficiency, finding applications in diverse fields such as search and rescue, exploration, and manufacturing (Iouguina et al., 2014; Pathak, 2019).

Fig. 1.3: Spider Inspired Roboticshttps://www.flickr.com/photos/
drewwith/3288999994/

Beyond functionality, bio-inspiration often results in aesthetically pleasing and innovative designs. Nature's designs, not just utilitarian but also elegant and beautiful, provide a source of inspiration that can have aesthetic and marketing benefits for engineered products. (Bensaude-Vincent, n.d.-a; Iouguina et al., 2014; Kennedy et al., 2015c). The collaborative nature of bio-inspired design is noteworthy. Encouraging interdisciplinary collaboration between engineers, biologists, and other experts fosters cross-disciplinary innovation. This holistic approach to problem-solving reflects the interconnectedness of various fields, leading to well-rounded and innovative solutions (Kennedy et al., 2015c). In the face of escalating environmental challenges, bio-inspired engineering emerges as a beacon of hope. Offering sustainable and eco-friendly solutions, it has the potential to mitigate the impact of human activities on the planet, contributing to a more resilient and sustainable future for society.

Summary

This chapter presents a comprehensive overview of bio-inspiration, starting with its definition as a design approach drawing from nature to address human challenges sustainably. It then delves into the historical context, tracing bio-inspired design from ancient civilizations to modern interdisciplinary practices, highlighting key milestones such as the emergence of biomimetics and NASA's Biomimetics program. The text further emphasizes the importance of bio-inspiration in engineering, detailing its role in deriving efficient, sustainable solutions across various domains, including materials science, robotics, and medicine. It discusses the benefits of bio-inspired engineering, such as resource efficiency, sustainability, and enhanced resilience, while also acknowledging its aesthetic and collaborative aspects. Overall, the summary encapsulates the evolution, significance, and interdisciplinary applications of bio-inspiration in fostering innovation and addressing environmental challenges.

References

Bensaude-Vincent, B. (n.d.-a). Bio-Informed Emerging Technologies and Their Relation to the Sustainability Aims of Biomimicry. *Environmental Values*.

Croasdell, V. L. (2019). *English: Sisters Aurora and Autumn Siegel, the youngest members of the MRISAR R&D team, began their apprenticeship in Robotics as preschoolers. The public use robotic exhibits they create for museums and science centers around the world relate to STEM and STEAM.* Own work. https://commons.wikimedia.org/wiki/File:Bio-inspired_7ft_robotic_spider.jpg

Iouguina, A., Dawson, J. W., Hallgrimsson, B., & Smart, G. (2014). *BIOLOGICALLY INFORMED DISCIPLINES: A COMPARATIVE ANALYSIS*

OF BIONICS, BIOMIMETICS, BIOMIMICRY, AND BIO-INSPIRATION AMONG OTHERS. 9(3).

Kennedy, E., Fecheyr-Lippens, D., Hsiung, B.-K., Niewiarowski, P. H., & Kolodziej, M. (2015a). *Biomimicry: A Path to Sustainable Innovation. 31(3).*

Lebdioui, A. (2022). Nature-inspired innovation policy: Biomimicry as a pathway to leverage biodiversity for economic development. *Ecological Economics.*

Pathak, S. (2019). *Biomimicry: (Innovation Inspired by Nature). 5(6).*

Taieb, A. H., & Amor, M. B. (n.d.). *Eco-design Design Strategy: Case Study Biomimicry Approach in Design Product Materials. 21(3).*

Withington, D. (2009). *"Elf Shelter" Rainwater Collector* [Photo]. https://www.flickr.com/photos/drewwith/3288999994/

CHAPTER II

Principles of Biomimicry

This chapter offers an in-depth exploration of biomimicry, elucidating its nature, key concepts, and real-world applications. Biomimicry, also known as bio-inspired design, emerges as a problem-solving approach that draws inspiration from nature's solutions honed over millions of years of evolution. It is characterized by interdisciplinary collaboration and a commitment to sustainability, aiming to emulate nature's efficiency and elegance in human-made innovations. The text delves into the fundamental principles and terminology of biomimicry, highlighting concepts such as biomimetic design, life's principles, and functional strategies employed by organisms. It emphasizes the importance of understanding the relationship between the form and function of natural systems and explores various examples of biomimicry in action, spanning industries like transportation, architecture, materials science, robotics, and sports. These examples showcase how biomimicry principles have been applied to create innovative and sustainable solutions, from Velcro inspired by burrs to the Shinkansen train influenced by a kingfisher's beak. The abstract encapsulates the essence of biomimicry as a transformative approach to innovation, harnessing nature's wisdom to address diverse challenges while fostering sustainability and efficiency in human endeavors.

2.1 The Nature of Biomimicry

Biomimicry, also known as bio-inspired design, is a problem-solving approach that draws inspiration from nature. Its nature can be summarized as follows: Biomimicry, as a discipline, involves a comprehensive exploration of natural systems, organisms, and processes to glean insights from the solutions developed through millions of years of evolution. It is rooted in the belief that nature inherently offers efficient and sustainable answers to many of the challenges confronting humanity. Going beyond passive observation, biomimicry actively seeks to emulate and imitate the principles, strategies, and designs found in nature. This could involve replicating the structure of a leaf to enhance solar panels or modeling transportation systems after the efficiency observed in ant colonies. (Aziz, n.d.; Bensaude-Vincent, n.d.-a; Lurie-Luke, 2014a)

An inherently interdisciplinary approach characterizes biomimicry, often fostering collaboration among biologists, engineers, designers, and experts from diverse fields. This interdisciplinary nature encourages a holistic perspective, enriching problem-solving processes. A fundamental tenet of biomimicry is the development of sustainable solutions. By replicating nature's strategies, biomimetic designs frequently yield more environmentally friendly and resource-efficiency products and systems, aligning seamlessly with the principles of sustainability and conservation. (Bensaude-Vincent, n.d.-a; Lurie-Luke, 2014a)

Biomimicry serves as a catalyst for innovative thinking and creative problem-solving, challenging conventional design and engineering approaches by seeking unconventional and elegant solutions from nature. Importantly, the process of biomimicry is not a one-time event but an ongoing journey of continuous learning and adaptation, recognizing nature as an ever-evolving source of inspiration.

Ethical considerations emerge prominently in the realm of biomimicry, prompting responsible reflections on the impact of bio-inspired technology on ecosystems and organisms. This calls for a thoughtful and ethical approach to the utilization of nature-inspired designs.

The applications of biomimicry extend across diverse fields, including architecture, transportation, materials science, medicine, and robotics, showcasing its versatility and applicability. Beyond practicality, biomimicry often results in aesthetically pleasing and visually striking designs. Nature's designs, in addition to their functionality, embody elegance and beauty.

Ultimately, the essence of biomimicry lies in making a positive impact on the world. By deriving lessons from nature and applying them to human challenges, biomimicry holds the potential to drive innovation, solve problems, and contribute to a more sustainable and harmonious relationship between human technology and the environment.

Biomimicry is a nature-inspired, interdisciplinary, and sustainable approach to problem-solving and innovation that has the potential to transform the way we design and engineer solutions for a wide range of challenges

Fig. 2.1: Pitcher Plant*pitcher plant—Google Search*

2.2 Key Concepts and Terminology

In the field of biomimicry, several key concepts and terminology help describe the principles and processes involved. Here are some of the fundamental terms and concepts:

Biomimicry encompasses the broad concept of leveraging insights from nature to address and solve various challenges faced by humanity. At its core

When the practical application of this concept, involving the creation of products, systems, or processes that either directly imitate or draw inspiration from biological model found in nature. The term "bio-inspiration" serves as a synonym for biomimicry, emphasizing the notion that it involves the derivation of inspiration from the intricate workings of biology.

Life's Principles, a fundamental aspect of biomimicry, constitute a set of sustainability principles extracted from the meticulous study of natural systems. These principles encompass vital aspects like adaptability, resource efficiency and closed-loop cycles, providing a foundational framework for sustainable design practices. (Amer, 2019; Bensaude-Vincent, n.d.-b)

Central to biomimetic design are the functional strategies employed by organisms in nature to adapt and survive in their environments. These strategies serve as specific models that can be emulated in the design of various products and systems, showcasing nature's efficiency and effectiveness in problem-solving.

The relationship between the form (structure) and function of natural systems is a pivotal consideration in biomimicry. Understanding how the structure of biological entities aligns with the functions they perform is crucial for successful biomimetic design, ensuring that the derived solutions are not only inspired by nature but also functionally effective.

Furthermore, biomimicry involves exploring the hierarchy of biological organization, spanning from the molecular level to the ecosystem level. This Comprehensive examination allows designers and scientists to understand how principles observed in natural systems can be applied across different scales, fostering a nuanced and adaptable approach to biomimetic design.

In essence, biomimicry encompasses a multifaceted exploration of nature's wisdom, involving not only the direct imitation of biological models but also a profound understanding of life's principles, functional strategies, the intricate relationship between form and function, and the hierarchical organization of biological systems. Through these lenses, biomimicry becomes a dynamic and versatile tool for innovative problem-solving, contributing to sustainable and effective solutions inspired by the brilliance of the natural world. (Amer, 2019; Bensaude-Vincent, n.d.-b; Lurie-Luke, 2014b)

Biomimicry delves into a spectrum of interconnected concepts, expanding our understanding of the intricate relationship between human innovation and the natural world. Analogous Structures highlight the incorporation of features in human-made products that closely mirror those found in nature, underlining the inspiration drawn from the efficiency of natural designs. Evolutionary Innovation embodies the recognition that designs in nature have evolved over time, tested and refined through the crucible of natural selection, providing a rich source of insights for human-created solutions. (Lakhtakia & Martín-Palma, 2013)

Ecosystem Services represent the invaluable benefits that natural ecosystems bestow upon humanity, including pollination, water purification, and climate regulation. Biomimicry seizes upon these services as models for sustainable solutions, mirroring the intricate balance observed in natural systems. Symbiosis, the exploration of mutualistic relationships in nature, serves as a wellspring of inspiration for the development of cooperative and mutually beneficial human systems. (Bensaude-Vincent, n.d.-b; Kennedy et al., 2015)

Morphology, the study of the form and structure of organisms, informs biomimetic design decisions, allowing for a deeper understanding of the aesthetic and functional aspects of natural designs. Adaptive Radiation, a natural process fostering the diversification and adaptation of organisms to new ecological niches, provides valuable insights for designing resilient and adaptable human systems.

Feedback Loops, mechanisms that organisms employ to regulate and optimize their functions, become a blueprint for engineering systems seeking efficiency and optimization. Closed-Loop Systems, observed in nature, exemplify the

recycling and reuse of materials, offering profound insights into sustainable material cycles within design practices.

Eco-mimicry, a subset of biomimicry, concentrates on emulating ecological systems and relationships to tackle complex environmental challenges. Emergent Properties, arising from the interactions of system components, inspire innovative solutions by demonstrating the power of synergy and interconnectedness. (Bensaude-Vincent, n.d.-a; Kennedy et al., 2015)

Bio-inspired Materials represent a cutting-edge frontier, designed to mimic the unique properties of natural materials. Biophilia, the intrinsic human connection to nature, becomes a focal point for biomimetic design, tapping into this connection to enhance well-being and productivity. (Bensaude-Vincent, n.d.-a)

Regenerative Design encapsulates principles that seek to restore and enhance ecosystems while meeting human needs, drawing inspiration from the regenerative processes inherent in nature. These diverse concepts collectively underscore the depth and breadth of biomimicry, showcasing its potential to revolutionize innovation by learning, adapting, and applying nature's ingenious solutions across a wide spectrum of human endeavors. (Amer, 2019; Aziz, n.d.)

Understanding these key concepts and terminology is essential for effectively applying biomimicry principles in design and innovation processes.

Fig. 2.2.2: Leaf Inspired Rainwater CollectorSource: (*"Elf Shelter" Rainwater Collector | The Elf Shelter Is a Gia... | Flickr*, n.d.)

2.3 Biomimicry in Action

Biomimicry in action involves applying the principles of biomimicry to various real-world challenges. Here are some examples of how these principles have been put into practice:

Biomimicry manifests in various ingenious applications, each a testament to nature's influence on human innovation. Velcro, conceived by Swiss engineer George de Mestral in the 1940s, owes its existence to the observation of burrs clinging to his dog's fur. This simple yet effective fastening system mimics the tiny hooks and loops on burrs, showcasing how nature's mechanisms can inspire practical solutions. (Bensaude-Vincent, n.d.-b; Kennedy et al., 2015)

The design of Japan's Shinkansen, the iconic bullet train, drew inspiration from the kingfisher's beak. Engineers emulated the streamlined shape of the bird's beak to reduce noise and resistance, resulting in a quieter and more efficient train. This adaptation illustrates how nature's efficiency can be harnessed to enhance technological designs. (Aziz, n.d.; Bensaude-Vincent, n.d.-b)

Biomimetic architecture takes cues from nature to optimize building performance. The Eastgate Centre in Zimbabwe, for instance, mimics termite mounds, incorporating passive cooling techniques inspired by termite ventilation systems. This architectural innovation not only reduces energy consumption but also maintains a comfortable interior climate, showcasing the potential for sustainable building practices.

In the realm of sports, swimsuit design takes inspiration from sharkskin to improve performance. Swimsuits featuring a texture inspired by sharkskin reduce drag in the water, a feature highly sought after by competitive swimmers striving for enhanced efficiency. This application underscores how biomimicry can lead to innovations that directly impact athletic performance. (Benyus, 1997; Lurie-Luke, 2014a)

The Lotus Effect serves as inspiration for self-cleaning surfaces used in architectural coatings and clothing. By emulating the lotus leaf's ability to repel water and contaminants, these surfaces demonstrate nature's prowess in maintaining cleanliness. This application not only enhances the durability of materials but also contributes to the development of low-maintenance and sustainable products. (Bensaude-Vincent, n.d.-a; Benyus, 1997)

These real-world examples highlight the versatility and practicality of biomimicry, showcasing its potential to revolutionize diverse industries. From fastening systems to transportation design, architectural solutions, sports apparel, and materials science, biomimicry continues to demonstrate its capacity to shape a more efficient, sustainable, and technologically advanced future by learning from the brilliance of the natural world.

Fig 2.3.1: The Lotus EffectSource: (*Lotus Flower Free Image | Peakpx,* n.d.)

Biomimicry emerges as a transformative force across various fields, each application showcasing the ingenious integration of nature-inspired principles into human innovation. Wind turbine blade design draws inspiration from the adapted flippers of humpback whales, minimizing turbulence and increasing lift to enhance the efficiency of wind energy production. This bio-inspired approach not only aligns with sustainability goals but also reflects the potential for more effective renewable energy solutions. (Benyus, 1997; Kennedy et al., 2015)

Biomimetic robotics takes center stage with creations like Boston Dynamics' Cheetah robot, mirroring the running mechanics of cheetahs. This bio-inspired locomotion brings forth remarkable agility, holding promise for applications such as search and rescue. The melding of robotics and nature's designs exemplifies the potential for creating machines with enhanced capabilities, driven by insights from the natural world.

In the context of materials science, the development of self-healing materials draws inspiration from the regenerative properties of human skin.

This innovation has far-reaching implications, potentially extending the lifespan of various products and contributing to a reduction in waste, epitomizing the harmonious fusion of biology and technology.

Gecko-inspired adhesives, modeled after the adhesive properties of gecko feet, showcase remarkable versatility. These adhesives adhere to various surfaces without leaving a residue, finding applications in robotics and industry. The mimicry of nature's adhesive mechanisms provides a blueprint for creating efficient and residue-free bonding solutions.

Biophilic design introduces principles of biophilia into interior and architectural design, aiming to create spaces that foster a profound connection between individuals and nature. This approach not only enhances the aesthetics of spaces but also leads to tangible benefits such as improved well-being, productivity, and creativity, emphasizing the symbiotic relationship between humans and the natural environment. (Benyus, 1997; Lakhtakia & Martín-Palma, 2013)

Fig 2.3.2: The Eastgate CentreSource: (Brown, 2018)

These examples illustrate how biomimicry principles have been successfully applied to solve a wide range of challenges, from improving product efficiency and sustainability to creating innovative designs and technologies that are inspired by the wisdom of nature. Biomimicry continues to be a source of inspiration for addressing complex problems in various fields.

Summary

This chapter offers an in-depth exploration of biomimicry, elucidating its nature, key concepts, and real-world applications. Biomimicry, also known as bio-inspired design, emerges as a problem-solving approach that draws inspiration from nature's solutions honed over millions of years of evolution. It is characterized by interdisciplinary collaboration and a commitment to sustainability, aiming to emulate nature's efficiency and elegance in human-made innovations. The text delves into the fundamental principles and terminology of biomimicry, highlighting concepts such as biomimetic design, life's principles, and functional strategies employed by organisms. It emphasizes the importance of understanding the relationship between the form and function of natural systems and explores various examples of biomimicry in action, spanning industries like transportation, architecture, materials science, robotics, and sports. These examples showcase how biomimicry principles have been applied to create innovative and sustainable solutions, from Velcro inspired by burrs to the Shinkansen train influenced by a kingfisher's beak. The abstract encapsulates the essence of biomimicry as a transformative approach to innovation, harnessing nature's wisdom to address diverse challenges while fostering sustainability and efficiency in human endeavors.

References

Amer, N. (2019). Biomimetic Approach in Architectural Education: Case study of 'Biomimicry in Architecture' Course. *Ain Shams Engineering Journal.*

Aziz, M. S. (n.d.). *Biomimicry as an approach for bio-inspired structure with the aid of computation.*

Bensaude-Vincent, B. (n.d.-a). Bio-Informed Emerging Technologies and Their Relation to the Sustainability Aims of Biomimicry. *Environmental Values*.

Bensaude-Vincent, B. (n.d.-b). Bio-Informed Emerging Technologies and Their Relation to the Sustainability Aims of Biomimicry. *Environmental Values*.

Benyus, J. (1997). Innovation inspired by nature: Biomimicry. *New York: William Morrow & Co.* https://eduscol.education.fr/sti/sites/ eduscol.education.fr.sti/files/ressources/techniques/1005/ 1005-172-p9.pdf

Brown, E. (2018). *East Gate and Tanybont Arch—Greengate Street, Caernarfon* [Photo]. https://www.flickr.com/photos/ell-r-brown/ 41901919184/

"Elf Shelter" Rainwater Collector | The Elf Shelter is a gia... | Flickr. (n.d.). Retrieved May 9, 2024, from https://www.flickr.com/photos/drewwith/ 3289000002

Kennedy, E., Fecheyr-Lippens, D., Hsiung, B.-K., Niewiarowski, P. H., & Kolodziej, M. (2015). Biomimicry: A path to sustainable innovation. *Design Issues*, *31*(3), 66–73. https://direct.mit.edu/desi/article-abstract/31/3/ 66/69195

Lakhtakia, A., & Martín-Palma, R. J. (2013). *Engineered biomimicry.* Newnes. https://books.google.com/books?hl=en&lr=&id=-

Lotus flower free image | Peakpx. (n.d.). Retrieved May 9, 2024, from https://www.peakpx.com/434864/lotus-flower

Lurie-Luke, E. (2014a). Product and technology innovation: What can biomimicry inspire? *Biotechnology Advances*, *32*(8), 1494–1505. https://www.sciencedirect.com/science/article/pii/S0734975014001517

Lurie-Luke, E. (2014b). Product and technology innovation: What can biomimicry inspire? *Biotechnology Advances*, *32*(8), 1494–1505. https://www.sciencedirect.com/science/article/pii/S0734975014001517

pitcher plant—Google Search. (n.d.). Retrieved May 9, 2024, from https://timelessmoon.getarchive.net/media/flower-plant-carnivore- nature-landscapes-15880a

CHAPTER III

Biological Inspirations

This chapter offers an exploration of biological organisms for inspiration across various disciplines, including technology, design, and medicine, is a fundamental aspect of biomimicry. This process involves studying the characteristics, behaviors, and adaptations of living creatures to derive insights for innovative solutions. Examples range from bio-inspired engineering designs like Velcro to the development of biopharmaceuticals derived from natural compounds. These inspirations lead to sustainable, efficient, and creative solutions across diverse fields. Moreover, biomimicry extends to a wide array of organisms, including plants, animals, and microorganisms, each offering unique features and adaptations that spark innovation. From the aerodynamics of bird flight influencing airplane design to the antibacterial properties of certain plants inspiring hygienic materials, nature serves as a wellspring of inspiration for solving complex challenges in human endeavors. These endeavors showcase the transformative potential of biomimicry in shaping advancements across disciplines, bridging the gap between natural ingenuity and human innovation.

3.1 Study of Natural Organisms

Studying natural organisms for biological inspiration is a common approach in various fields, including technology, design, and medicine. It involves observing and analyzing the characteristics, behaviors, and adaptations of living creatures to derive insights and innovative solutions. Examples include biomimicry in engineering, where designs are inspired by nature (e.g., Velcro modeled after burrs) and biopharmaceuticals developed from compounds found in organisms (e.g., antibiotics from fungi). This approach can lead to sustainable, efficient, and creative solutions.

Fig. 3.1.1: Macroscopic View of Velcro

3.2 Plants, Animals, and Microorganisms

Biological inspiration can be drawn from a wide range of natural organisms, including plants, animals, and microorganisms. Here are some examples of how each category can provide inspiration in various fields:

Biomimicry manifests across a diverse spectrum, drawing inspiration from nature's myriad designs and mechanisms to revolutionize technology, design, and biology. The iconic invention of Velcro by Swiss engineer George de Mestral stemmed from observing how burrs adhere to clothing, providing a simple yet effective fastening system. In the realm of renewable energy, biomimicry has influenced the development of solar panels, mirroring the efficient process of photosynthesis in plants. Similarly, biomimetic materials inspired by the water-repellent properties of lotus leaves are in the works, promising self-cleaning surfaces with applications ranging from architecture to everyday products. (Bensaude-Vincent, n.d.-a)

In the animal kingdom, the aerodynamics of bird flight have influenced the design of airplanes and drones, exemplifying how biomimicry can enhance human-made flight systems. Sharkskin-inspired swimsuits and materials, designed to reduce drag in water, showcase the application of nature's efficiency in sports apparel. Additionally, adhesive materials inspired by gecko feet have been created, finding applications in innovative technologies like wall-climbing robots. (Bensaude-Vincent, n.d.-a; Kennedy et al., 2015)

Microorganisms, often overlooked but crucial in biomimicry, have contributed significantly to advancements in technology and biology. Many antibiotics, including the groundbreaking penicillin from Penicillium fungi, find their roots in microorganisms. In environmental applications, microbes are harnessed for bioremediation, playing a key role in cleaning up oil spills and contaminated environments. Moreover, biological sensors utilizing microorganisms have been developed to detect pollutants or pathogens in water and soil, showcasing the potential for biomimicry to address environmental challenges. (Amer, 2019; Kennedy et al., 2015)

The study of these natural organisms provides a rich source of insights, influencing innovations that span various disciplines. Whether it's the efficiency of Velcro, the energy capture of solar panels, the aerodynamics inspired by bird flight, the streamlined design from sharkskin, the adhesive capabilities of gecko feet, or the medicinal properties of antibiotics, biomimicry continues to shape advancements across diverse fields. There's a specific area within these categories you're keen to explore further.

Fig. 3.2.1: Starfish(*Marine Life,Starfish,Mussels,Ocean Life,Ocean - Free Image from Needpix.Com, n.d.*)

3.3 Unique Features and Adaptations

Biological organisms, with their diverse adaptations, have served as a wellspring of inspiration for innovations across multiple domains. A striking example lies in the realm of camouflage and mimicry, where the intricate patterns observed in animals like chameleons or cuttlefish have influenced the design of military uniforms and equipment. The incorporation of these patterns aims to enhance concealment and disguise in various environments, showcasing how nature's strategies are leveraged to improve human technologies. (Amer, 2019; Bensaude-Vincent, n.d.-b; Kennedy et al., 2015)

Furthermore, the concept of mimicry extends into the field of robotics. Here, machines are designed to imitate the appearance and behavior of animals, allowing them to seamlessly blend into natural environments. This biomimetic approach in robotics not only enhances the machines' ability to navigate and operate in different surroundings but also demonstrates the potential for creating highly adaptable and inconspicuous robotic systems. In essence, the study of camouflage and mimicry in biological organisms has led to practical applications that optimize human technologies for improved functionality and efficiency. (Amer, 2019; Aziz, n.d.; Lakhtakia & Martín-Palma, 2013a; Lurie-Luke, 2014)

Fig.3.3.1: Chameleon(*Chameleon,Camouflage,Exotic,Close,Nature - Free Image from Needpix.Com*, n.d.)

Nature's ingenuity extends to various biological phenomena, sparking innovations that redefine technological landscapes. Bioluminescence, observed in organisms like fireflies and deep-sea creatures, has inspired the development of energy-efficient lighting systems. The enchanting glow of these creatures serves as a blueprint for creating sustainable illumination, contributing to eco-friendly and visually captivating lighting solutions.

Spider silk, renowned for its remarkable strength and lightweight nature, serves as a muse for bioengineered materials. These materials find applications in diverse fields, from textiles to medical devices and even bulletproof clothing. By harnessing the properties of spider silk, researchers and engineers pave the way for the creation of advanced materials that combine strength with flexibility. (Benyus, 1997; Lakhtakia & Martín-Palma, 2013a)

The echolocation abilities of bats and dolphins have influenced the development of sonar technology. This bio-inspired approach has been instrumental in the design of submarines and navigation systems, showcasing how nature's navigation strategies can be emulated to enhance human technologies. (Benyus, 1997)

Photosynthesis, the process by which plants convert sunlight into energy, has spurred the creation of bio-inspired energy systems. Notably, solar panels mimic this natural phenomenon to generate electricity sustainably. This application of biomimicry not only aligns with the principles of renewable energy but also underscores the potential of learning from nature's intricate processes. (Bensaude-Vincent, n.d.-b; Benyus, 1997)

Materials with antibacterial properties have been developed by drawing inspiration from the natural antimicrobial surfaces of certain plants and insects. This bio-inspired approach has implications for creating hygienic and self-cleaning surfaces in various settings, contributing to improved public health.

Exploring the concept of hibernation in animals has led to research aimed at inducing hibernation-like states in humans. This bio-inspired endeavor holds promise for long-duration space travel and medical applications, showcasing the potential for integrating nature's adaptive mechanisms into human biology.

Honeycomb structures observed in beehives and bone tissues have influenced architectural designs, aircraft components, and lightweight materials.

By emulating the efficiency of these natural structures, engineers create innovative solutions that balance strength, durability, and weight, contributing to advancements in architecture and aerospace engineering. In essence, these examples highlight the breadth of biomimicry's impact, showcasing how insights from nature's designs continue to shape and elevate human technology across a diverse array of fields. (Kennedy et al., 2015; Lakhtakia & Martín-Palma, 2013b)

Fig. 3.3.2: Macroscopic View of HoneycombSource: (Ltd, n.d.)

These unique features and adaptations found in the natural world often serve as a source of inspiration for innovative solutions in various fields.

Summary

This chapter explores the practice of drawing inspiration from biological organisms for innovation, known as biomimicry, across various fields such as technology, design, and medicine. It discusses how observing the characteristics, behaviors, and adaptations of living organisms can lead to creative and sustainable solutions to complex challenges. Examples include the development of Velcro inspired by burrs and the creation of biopharmaceuticals derived from natural compounds. The study highlights the wide range of organisms, including plants, animals, and microorganisms, that serve as sources of inspiration for innovative designs and technologies. From the aerodynamics of bird flight influencing aircraft design to the antibacterial properties of certain plants inspiring hygienic materials, biomimicry offers a transformative approach to problem-solving by leveraging nature's ingenuity.

References

Https://www.flickr.com/photos/paul_garland/3581676336—Yahoo Search Results. (n.d.). Retrieved May 3, 2024, from

https://search.yahoo.com/search?

https://shorturl.at/4rcPq

Amer, N. (2019). Biomimetic Approach in Architectural Education: Case study of 'Biomimicry in Architecture' Course. *Ain Shams Engineering Journal.*

Aziz, M. S. (n.d.). *Biomimicry as an approach for bio-inspired structure with the aid of computation.*

Bensaude-Vincent, B. (n.d.-a). Bio-Informed Emerging Technologies and Their Relation to the Sustainability Aims of Biomimicry. *Environmental Values.*

Bensaude-Vincent, B. (n.d.-b). Bio-Informed Emerging Technologies and Their Relation to the Sustainability Aims of Biomimicry. *Environmental Values.*

Benyus, J. M. (1997). *Biomimicry: Innovation inspired by nature.* Morrow New York.

https://www.academia.edu/download/5239337/biomimicry-innovation-inspired-by-nature.pdf

Chameleon,camouflage,exotic,close,nature—Free image from needpix.com. (n.d.). Retrieved

May 3, 2024, from https://www.needpix.com/photo/1079449/chameleon-camouflage-exotic-close-nature-stalk-eyed-terrarium-animals-creature-reptile

Kennedy, E., Fecheyr-Lippens, D., Hsiung, B.-K., Niewiarowski, P. H., & Kolodziej, M. (2015). Biomimicry: A path to sustainable innovation. *Design Issues, 31*(3), 66–73.

https://direct.mit.edu/desi/article-abstract/31/3/66/69195

Lakhtakia, A., & Martín-Palma, R. J. (2013a). *Engineered biomimicry.* Newnes.

https://shorturl.at/19ByB

Lakhtakia, A., & Martín-Palma, R. J. (2013b). *Engineered biomimicry.* Newnes.

https://shorturl.at/k9RP7

Ltd, B. (n.d.). *Honey Bee Hive Honeycomb Background Free Stock Photo—Public Domain Pictures.* Retrieved May 3, 2024, from
https://www.publicdomainpictures.net/en/view-image.php?image=358296&picture=honey-bee-hive-honeycomb-background

Lurie-Luke, E. (2014). Product and technology innovation: What can biomimicry inspire? *Biotechnology Advances, 32*(8), 1494–1505.
https://www.sciencedirect.com/science/article/pii/S0734975014001517

Marine life,starfish,mussels,ocean life,ocean—Free image from needpix.com. (n.d.). Retrieved
May 3, 2024, from https://www.needpix.com/photo/1550283/marinelife-starfish-mussels-oceanlife-ocean-sealife-orangecolor-shells-beach

CHAPTER IV

Bio-Inspired Material

This chapter offers an exploration of the practice of drawing inspiration from biological organisms for innovation, known as biomimicry, across various fields such as technology, design, and medicine. It discusses how observing the characteristics, behaviors, and adaptations of living organisms can lead to creative and sustainable solutions to complex challenges. Examples include the development of Velcro inspired by burrs and the creation of biopharmaceuticals derived from natural compounds. The study highlights the wide range of organisms, including plants, animals, and microorganisms, that serve as sources of inspiration for innovative designs and technologies. From the aerodynamics of bird flight influencing aircraft design to the antibacterial properties of certain plants inspiring hygienic materials, biomimicry offers a transformative approach to problem-solving by leveraging nature's ingenuity.

4.1 Materials Science and Engineering

Bio-inspired materials are a fascinating area of materials science and engineering where the properties and structures of biological organisms serve as inspiration for creating new materials with unique characteristics. Here are some examples of bio-inspired materials:

The remarkable qualities of spider silk, known for its lightweight nature combined with exceptional strength, have spurred the creation of synthetic materials that replicate its intricate structure. These biomimetic materials find applications across diverse fields, from textiles and medical sutures to advanced body armor. In textiles, the spider silk-inspired materials offer a unique combination of strength and flexibility, contributing to the creation of durable yet lightweight fabrics. In the medical field, the use of biomimetic spider silk has resulted in improved sutures, providing enhanced strength and minimizing tissue damage during surgical procedures. Additionally, the development of body armor incorporating synthetic spider silk highlights its potential to offer robust protection while maintaining a lightweight profile. The biomimetic approach, drawing inspiration from nature's design, showcases how spider silk's exceptional properties can be harnessed to address various challenges and innovate within different industries.

Fig. 4.4.1: Spider Web(*Free Images*, n.d.)

Nature's design principles have inspired a range of innovative materials with diverse applications. The lotus-leaf effect, characterized by water-repellent properties, has led to the creation of superhydrophobic materials. These are employed in self-cleaning surfaces and water-repellent clothing, showcasing the potential for biomimicry to enhance daily life. Drawing inspiration from bones, materials combining strength and lightweight characteristics have been developed for aerospace and automotive applications, contributing to advancements in structural design. Gecko-inspired adhesives, mimicking the reptile's ability to adhere to surfaces, find use in climbing robots and medical bandages, demonstrating the versatility of biomimetic solutions. Biomimetic composites, integrating natural elements like wood fibers, offer sustainable alternatives for construction and packaging materials, aligning with eco-friendly practices. Nacre, found in seashells, has inspired the creation of strong and flexible composite materials suitable for construction and protective gear. Moreover, the development of bio-inspired polymers replicating the properties of biological tissues, such as elasticity and self-healing capabilities, has paved the way for innovative applications in both medical and engineering domains. These biomimetic materials exemplify the fusion of nature-inspired design and technological innovation, showcasing the vast potential of bio-inspired solutions across various industries. (Bilici et al., n.d.; de Pauw, n.d.; Perera & Coppens, n.d.)

Fig. 4.1.2: The Lotus Effect(*Lotus Flower Free Image | Peakpx, n.d.*)

Biomineralization, a process observed in organisms like mollusks, has become a wellspring of inspiration for the development of materials with controlled crystal growth. This biomimetic approach involves mimicking the natural mechanisms by which living organisms produce minerals with precision. The insights gained from biomineralization have been particularly impactful in the fields of optics and electronics. By replicating the intricate control over crystal formation seen in nature, scientists and engineers have sought to enhance the performance and functionality of materials used in optical devices and electronic components. This bio-inspired strategy not only harnesses the efficiency of biological processes but also opens up new avenues for creating advanced materials that can push the boundaries of technology in ways that conventional approaches might not achieve. The study and application of biomineralization highlights the potential for biomimicry to drive innovation across diverse scientific and technological disciplines. ("Biomimicry, an Approach, for Energy Efficient Building Skin Design," 2016; de Pauw, n.d.)

Bio-inspired materials often combine the best of nature's design with the benefits of advanced engineering to create innovative solutions in a wide range of industries, from aerospace to healthcare.

4.2 Biomimetic Material Development

Biomimetic material development is a field that focuses on creating materials inspired by the structures and properties of biological organisms. The goal is to design and engineer materials that replicate or adapt the remarkable features found in nature. Here's how biomimetic material development works:

The process of developing biomimetic materials is a meticulously structured journey that begins with a profound understanding of nature's intricacies. Researchers embark on a comprehensive study of biological organisms, tissues, or processes they intend to mimic, delving into the detailed analysis of structures, properties, and functions. Following this understanding, the selection of suitable materials becomes paramount, with scientists choosing from an array of options like synthetic polymers, metals, ceramics, or composites to replicate the desired biological feature. (Perera & Coppens, n.d.; Reed et al., 2009)

With materials in hand, engineers move on to the design and fabrication phase, utilizing their acquired knowledge to create materials that faithfully mimic the chosen biological feature. This often involves the utilization of cutting-edge manufacturing techniques such as 3D printing or nanofabrication, pushing the boundaries of innovation in material design.

The journey doesn't end with creation; rigorous testing and optimization are crucial steps in ensuring that the newly developed biomimetic materials align with the desired properties. Iterative design processes come into play, allowing engineers to fine-tune the material's performance to meet specific criteria and standards.

Upon successful development, these biomimetic materials find diverse applications across various industries, ranging from aerospace to healthcare and construction. The transformative potential of biomimicry unfolds as artificial muscles, inspired by the contraction of human muscles, become instrumental in robotics and prosthetics. Similarly, bio-inspired adhesives that mimic the remarkable adhesion properties of gecko feet offer innovative solutions in industries requiring efficient and residue-free bonding.

This structured process, from understanding nature's designs to the practical application of biomimetic materials, exemplifies the intricate interplay between biological inspiration and human ingenuity. As researchers continue to unlock the secrets of nature, the potential for developing advanced biomimetic materials promises to reshape industries and pave the way for innovative solutions that seamlessly integrate with the natural world.

Fig. 4.2.1: The Burdock Plant(*Velcro Czepy, Burdock, the Ball, Flower, Dried - Free Image from Needpix.Com*, n.d.)

- Self-healing materials inspired by the regenerative abilities of biological tissues. (Bilici et al., n.d.)
- Bio-inspired composites that replicate the structure of natural materials like nacre (mother-of-pearl). (de Pauw, n.d.)

Biomimetic material development is a multidisciplinary field that combines biology, materials science, and engineering to create innovative solutions with a wide range of practical applications.

4.3 Case Studies in Material Innovation

There have been several remarkable case studies in material innovation inspired by biology. Here are a few notable examples:

Bio-Inspired Adhesives - Gecko Tape:

Taking inspiration from the adhesive properties of gecko feet, biomimicry has led to the innovation of gecko-inspired adhesive tapes and materials. These materials exhibit the remarkable ability to adhere to various surfaces without leaving any residue behind. Bio-inspired adhesives find applications in diverse fields, including robotics, medical devices, and beyond. The development of gecko-inspired adhesives introduces a new dimension to adhesion technology, offering solutions that are versatile, residue-free, and adaptable to different surfaces. (Amer, 2019; de Pauw, n.d.; Taieb & Amor, n.d.)

Bulletproof Clothing - Spider Silk:

Spider silk's extraordinary strength and lightweight properties have inspired the creation of synthetic materials for use in bulletproof clothing and lightweight body armor. Innovators have successfully replicated the characteristics of spider silk, producing materials that provide exceptional protection without the added weight often associated with traditional bulletproof materials. This biomimetic approach to developing protective clothing showcases the potential for lightweight and highly effective solutions in personal safety and defense. (de Pauw, n.d., n.d.; Taieb & Amor, n.d.)

Self-Healing Materials - Biological Tissues:

Drawing inspiration from the regenerative abilities of biological tissues, biomimetic self-healing materials have been developed. These materials possess the unique capability to repair damage autonomously without requiring external intervention. The application of self-healing materials extends across industries, including aerospace, construction, and consumer products. By emulating the regenerative mechanisms found in living organisms, these materials contribute to enhanced durability and longevity in various products and structures. (Bilici et al., n.d.)

Photoactive Materials - Photosynthesis:

Inspired by the intricate process of photosynthesis in plants, bio-inspired photoactive materials have been engineered to replicate the conversion of sunlight into energy. These materials find crucial applications in solar panels and renewable energy technologies. By mimicking nature's ability to harness solar energy efficiently, photoactive materials contribute to advancements in sustainable energy production, offering innovative solutions for capturing and utilizing sunlight. (Kennedy et al., 2015; Taieb & Amor, n.d.)

Aerogels - Dragonfly Wings:

Taking cues from the lightweight and robust structure of dragonfly wings, biomimicry has led to the development of aerogels. These materials emulate the unique properties of dragonfly wings, combining extreme lightness with strength. Biomimetic aerogels find applications in various fields, including insulation, oil spill cleanup, and lightweight structural components. Inspired by the natural design of dragonfly wings, these aerogels showcase the potential for innovative materials that balance strength and weight in diverse practical applications. (de Pauw, n.d.; Taieb & Amor, n.d.)

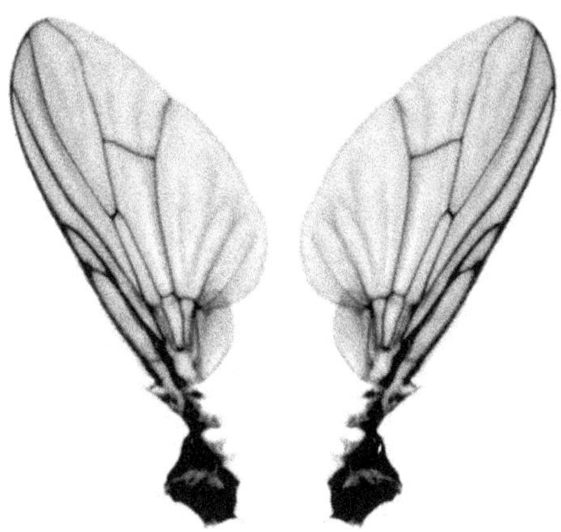

Fig. 4.3.1: The Dragonfly WingsSource: (Heisey, 2016)

Aerogels, drawing inspiration from the intricate structure of dragonfly wings, represent a groundbreaking innovation in material science. These ultra-light materials, modeled after the design principles found in nature, have found diverse applications in areas such as insulation, aerospace engineering, and oil spill cleanup. The biomimetic approach to creating aerogels leverages the lightweight yet robust structure of dragonfly wings, showcasing the potential for developing materials that combine strength with extreme lightness. This innovation not only provides solutions for improving energy efficiency and aerospace technologies but also addresses environmental challenges, exemplified by their use in oil spill cleanup operations. (Bilici et al., n.d.-b; "Biomimicry, an Approach, for Energy Efficient Building Skin Design," 2016)

Self-cleaning surfaces, inspired by the water-repellent properties of lotus leaves, represent another notable application of biomimicry. The innovative concept involves mimicking the micro- and nanostructures on lotus leaves that repel water and prevent the adhesion of contaminants. The resulting biomimetic materials, which exhibit self-cleaning properties, have been applied to surfaces in various industries, including architecture and consumer products. The inspiration from lotus leaves provides a sustainable and efficient solution for maintaining cleanliness and reducing the need for chemical cleaning agents. This biomimetic approach to self-cleaning surfaces aligns with eco-friendly practices, contributing to both technological advancements and environmental sustainability. ("Ecological Sustainability, Nature –Inspired, Architecture, Zero- Waste Systems, Regenerative Design," 2018)

Fig. 4.3.2: The Lotus Effect(*Lotus Flower Free Image | Peakpx*, n.d.)

The innovation of self-cleaning materials, inspired by the water-repellent properties of lotus leaves, marks a significant stride in biomimicry. These materials, with surfaces resembling those of lotus leaves, actively repel water and dirt, presenting a revolutionary solution to the challenges of maintaining cleanliness across various applications. This biomimetic approach has led to the development of coatings for buildings, textiles, and automotive surfaces, where the self-cleaning properties reduce the need for frequent cleaning and contribute to the longevity of surfaces. By emulating the natural features of lotus leaves, these biomimetic materials showcase a sustainable and eco-friendly alternative, aligning with the growing emphasis on environmentally conscious technologies. ("Ecological Sustainability, Nature –Inspired, Architecture, Zero- Waste Systems, Regenerative Design," 2018; Ilieva et al., 2022)

The biomimicry of nacre, the iridescent inner layer of seashells, has inspired the innovation of composite materials with exceptional strength and flexibility. Nacre's layered structure, which provides both resilience and flexibility, has been replicated in composite materials. These biomimetic composites find applications in critical areas such as aircraft components and protective gear, where a combination of strength and durability is paramount. The mimicry of nacre's structure not only enhances the mechanical properties of these materials but also demonstrates how nature's design principles can be harnessed to create advanced, high-performance synthetic materials for various industrial uses.

Fig. 4.3.3: Picture Of Bio-PlaneSource: (Caribb, 2006)

These case studies illustrate how insights from nature can lead to the development ofinnovative materials with a wide range of applications. Biomimetic material innovation continues to be an exciting and growing field, with researchers drawing inspiration from the natural world to solve complex engineering challenges

Summary:

This chapter delves into the realm of bio-inspired materials, a captivating field within materials science and engineering that draws inspiration from nature's designs to create innovative materials with remarkable properties. It explores examples such as synthetic materials inspired by spider silk's strength and lightweight nature, biomimetic composites replicating the structure of natural materials like nacre, and bio-inspired polymers mimicking the properties of biological tissues. The process of biomimetic material development involves a meticulous journey, starting with a deep understanding of biological organisms, followed by material selection, design, fabrication, testing, and optimization. Case studies highlight remarkable innovations, including bio-inspired adhesives based on gecko feet, bulletproof clothing derived from spider silk, and self-healing materials inspired by biological tissues. These examples showcase how biomimicry leads to the creation of materials with diverse applications, from aerospace to healthcare, and illustrate the transformative potential of nature-inspired design principles in addressing complex engineering challenges while fostering sustainability and innovation.

References:

Amer, N. (2019). Biomimetic Approach in Architectural Education: Case study of 'Biomimicry in Architecture' Course. *Ain Shams Engineering Journal.*

Bilici, S. C., Küpeli, M. A., & Guzey, S. S. (n.d *Inspired by nature: An engineering design-based biomimicry activity.*

de Pauw, I. C. (n.d.). Comparing Biomimicry and Cradle to Cradle with Ecodesign: A Case Study of Student Design Projects. *MANUS CRIP T.*

Ecological sustainability, Nature –inspired, Architecture, Zero- waste systems, Regenerative design. (2018). *Architecture Research.*

Ilieva, L., Ursano, I., Traista, L., Hoffmann, B., & Dahy, H. (2022). *Biomimicry as a Sustainable Design Methodology—Introducing the 'Biomimicry for Sustainability' Framework.*

Kennedy, E., Fecheyr-Lippens, D., Hsiung, B.-K., Niewiarowski, P. H., & Kolodziej, M. (2015). Biomimicry: A path to sustainable innovation. *Design Issues, 31*(3), 66–73. https://direct.mit.edu/desi/article-abstract/31/3/66/69195

Perera, A. S., & Coppens, M.-O. (n.d.). *Re-designing materials for biomedical applications: From biomimicry to nature-inspired chemical engineering.*

Reed, E. J., Klumb, L., Koobatian, M., & Viney, C. (2009). Biomimicry as a route to new materials: What kinds of lessons are useful? *New Materials.*

Taieb, A. H., & Amor, M. B. (n.d.). *Eco-design Design Strategy: Case Study Biomimicry Approach in Design Product Materials. 21*(3).

CHAPTER V

Biomimetic Structures and Design

This chapter offers an exploration of the application of biomimetic structures and design principles across various fields, including architecture, aerospace, and vehicle design. It highlights how drawing inspiration from nature leads to the creation of more sustainable, efficient, and innovative structures. Key concepts such as biophilic design, biomorphic architecture, and bionics are discussed, along with case studies illustrating the implementation of biomimetic approaches in real-world projects. Lessons from nature, including optimal shapes, lightweight materials, self-repair mechanisms, and efficient energy use, are analyzed for their relevance to human-made structures. The abstract encapsulates the transformative potential of biomimicry in shaping the future of design and technology, emphasizing sustainability, efficiency, and integration with natural ecosystems.

5.1 Architecture and Building Design

Biomimetic structures and design in architecture and building design draw inspiration from nature to create more sustainable, efficient, and aesthetically pleasing structures. Here are some key examples and concepts in this field:

Biophilic design represents a holistic approach to architecture, seamlessly blending the built environment with nature to enhance occupants' well-being. This innovative design philosophy embraces natural elements and patterns, creating spaces that resonate with the inherent human connection to nature. Practical implementations may involve the incorporation of features such as abundant natural lighting, the introduction of indoor plants, or the use of materials that replicate natural textures. By integrating these elements, biophilic design seeks to create environments that foster improved mental health, productivity, and overall satisfaction among occupants.

In a striking example of biomimicry, termite mound architecture draws inspiration from the intricate structures created by termites in Africa. The design of energy-efficient buildings is informed by the termites' natural ventilation systems. These biomimetic structures utilize passive cooling and heating principles, optimizing temperature regulation within the building. By mimicking the efficiency of termite mounds, architects and engineers contribute to sustainable practices in construction, reducing energy consumption and minimizing the environmental impact of the built environment. This intersection of biophilic design and biomimicry exemplifies the harmonious integration of natural principles into architectural innovation, creating spaces that not only prioritize human well-being but also demonstrate a sustainable and eco-friendly approach to building design. (Bilici et al., n.d.-a; E. B. Kennedy & Marting, 2016)

Biomorphic architecture embraces the concept of bio morphism, incorporating organic shapes and patterns into designs, resulting in visually striking and environmentally friendly structures. An exemplary manifestation is seen in the Eden Project's biomes, resembling giant soap bubbles. This innovative approach not only creates architecturally distinctive buildings but also underscores a commitment to sustainable design, demonstrating how bio morphism can redefine the aesthetics and ecological impact of architectural spaces. (Fiorentino & Montana-Hoyos, n.d.; Huh et al., 2012)

Bionics plays a pivotal role in shaping lightweight structures inspired by the efficiency found in nature. Architects and engineers draw insights from the natural world to design buildings with innovative materials and construction techniques. The Water Cube in Beijing stands as a testament to this approach, inspired by the lightweight and sturdy nature of soap bubbles. By translating these principles into architecture, professionals contribute to the development of structures that are both resilient and resource efficient. (Fiorentino & Montana-Hoyos, n.d.)

Green roofs and living walls represent a biophilic design strategy that integrates vegetation into building exteriors. Inspired by the lush greenery found in natural ecosystems, these designs offer benefits such as improved insulation, enhanced air quality, and enhanced aesthetics. This synthesis of nature and architecture not only transforms urban landscapes but also promotes sustainability and well-being. (E. B. Kennedy & Marting, 2016)

Sustainable materials and construction techniques are pivotal in biomimetic approaches to building design. Bamboo, known for its rapid growth and strength, stands out as a prime example of a biomimetic material frequently used in green construction. By emulating nature's efficiency and resilience, architects contribute to the development of eco-friendly and durable structures. (E. B. Kennedy & Marting, 2016)

Solar energy optimization draws inspiration from plants' ability to maximize sunlight exposure. The design of solar panels and sun-tracking systems takes cues from nature's efficiency, resulting in more effective energy production in buildings. This biomimetic approach not only enhances the sustainability of energy systems but also exemplifies how nature-inspired design principles can shape the future of renewable energy technologies. In essence, these examples showcase the diverse ways in which biomimetic principles are transforming architect, pushing the boundaries of creativity, sustainability, and efficiency in the built environment. (Araque et al., 2021; Bilici et al., n.d.-a, n.d.-b)

Fig. 5.1.1: Solar PannelSource: (*Free Images: Sky, Sun, Technology, Sunlight, Solar Panel, Solar Power, Solar Energy, Photovoltaics, Solar Panels 5184x3456 - - 708107 - Free Stock Photos - PxHere*, n.d.)

Aerodynamic Building Shapes: The aerodynamic shapes of birds and fish can inspire building designs that reduce wind resistance and improve energy efficiency. Biomimetic architecture and building design aim to create structures that harmonize with the environment, enhance human well-being, and reduce the environmental impact. These innovative approaches often result in visually appealing and energy-efficient buildings that take inspiration from the natural world.

5.2 Aerospace and Vehicle Design

Biomimetic structures and design have made significant contributions to the fields of aerospace and vehicle design, where engineers and designers draw inspiration from nature to create more efficient and aerodynamic vehicles. Here are some examples: The biomimicry of bird wings has profoundly influenced the design and innovation of aircraft wings, introducing advanced features that enhance performance and efficiency. Specifically, the incorporation of winglets, inspired by bird wing shapes, represents a significant breakthrough in aeronautical engineering. These winglets emulate the intricate curvature observed in bird wings and contribute to minimizing drag, consequently improving overall fuel efficiency in aircraft. Moreover, the study of owl wing structures has led to innovations in quieter aircraft designs, aiming to mitigate noise pollution. By understanding and replicating the unique features of bird wings, aircraft engineers have not only optimized aerodynamics but also addressed environmental concerns, showcasing the transformative impact of biomimicry in aviation technology. (Ilieva et al., 2022; E. B. Kennedy & Marting, 2016)

Fig. 5.2.1: A bird flying in the skyA bird flying in the sky with its wings spread

The field of aviation and aeronautics has extensively drawn inspiration from the remarkable agility and maneuverability observed in birds and insects. This influence is notably evident in the design of both drones and aircraft, which have been engineered to emulate the intricate flight capabilities of their natural counterparts. The capacity for these biomimetic vehicles to execute precise and agile maneuvers is particularly valuable in applications such as search and rescue missions and surveillance, where adaptability and swift movement are essential. (Ilieva et al., 2022; E. B. Kennedy & Marting, 2016)

Furthermore, biomimicry plays a pivotal role in the development of lightweight and robust materials for aircraft and spacecraft construction. By studying the structure of bones and other lightweight materials found in nature, engineers have successfully created materials that enhance the overall performance and fuel efficiency of aerospace technologies. This biomimetic approach addresses the dual objectives of reducing weight for increased efficiency while maintaining structural integrity. (Huh et al., 2012)

In the realm of renewable energy, biomimicry has influenced wind turbine blade design. By examining the shape and movement of humpback whale flippers, engineers have derived more efficient blade designs that optimize energy generation from wind power. This integration of biological insights into engineering solutions showcases the potential of biomimicry to revolutionize various aspects of aviation and sustainable energy production. (Ilieva et al., 2022).

Fig. 5.2.2: Turbine Blade, Humpback WhaleBilici, S. C., Küpeli, M. A., & Guzey, S. S.

The integration of biomimicry into aerospace and vehicle design has yielded innovations focused on improving efficiency, reducing energy consumption, enhancing maneuverability, and minimizing environmental impact. Drawing inspiration from the streamlined body shapes of fish, particularly tuna, engineers have designed submersibles and underwater vehicles that exhibit greater hydrodynamics, translating into improved energy efficiency beneath the water's surface.

Researchers have also looked to nature for insights into the unique hovering flight of bees. This understanding has led to the development of small drones and vehicles capable of hovering and maneuvering in confined spaces. This bio-inspired approach enhances adaptability, making these vehicles suitable for various applications, including surveillance and exploration.

In the quest for fuel efficiency in commercial aircraft, biomimicry has turned to the long-distance flight patterns of migratory birds. Insights from bird migration have informed the development of flight routing algorithms, contributing to more fuel-efficient routes and reduced environmental impact.

Another notable biomimetic application involves sharkskin-inspired coatings on ships and submarines. By mimicking the texture of sharkskin, these coatings aim to reduce drag and enhance efficiency in water-based transport.

Therefore, biomimetic designs in aerospace and vehicle development showcase a commitment to sustainable practices and innovative solutions. By emulating the efficiency of natural processes, engineers and designers strive to create vehicles that not only outperform their conventional counterparts but also contribute to a more sustainable and environmentally conscious transportation landscape. (Ilieva et al., 2022)

5.3 Lessons from Nature for Efficient Structures

Biomimetic structures and design offer valuable lessons from nature for creating efficient and innovative structures across various fields. Here are some key takeaways:

Nature serves as a profound source of inspiration for optimizing various aspects of human technology and design. One key area is the study of optimal shapes found in natural structures, which often minimize energy consumption. The streamlined forms observed in fish and birds, for instance, have become a blueprint for creating more efficient vehicle and aircraft designs. By emulating these shapes, engineers enhance the aerodynamics of vehicles, contributing to reduced energy consumption and improved overall performance. (Araque et al., 2021)

Another invaluable lesson from nature lies in lightweight materials. Many biological structures showcase the remarkable combination of being both lightweight and strong. Engineers leverage insights from materials like bone or spider silk to craft innovative materials for applications in aerospace, construction, and beyond. This biomimetic approach results in materials that offer a high strength-to-weight ratio, unlocking new possibilities for efficient and resilient structures. (Bilici et al., n.d.-a;

Verbrugghe et al., 2023)

Self-repair mechanisms, inherent in biological tissues, inspire biomimetic materials and designs with the ability to autonomously repair damage. This incorporation of self-repairing capabilities reduces maintenance requirements and extends the lifespan of structures, presenting a sustainable and cost-effective solution across various industries. (Bilici et al., n.d.-a; E. B. Kennedy & Marting, 2016)

Efficient energy use finds inspiration in observing how animals like ants or bees navigate their environments with remarkable efficiency. The development of algorithms and systems for optimizing energy use in buildings and transportation draws insights from nature's streamlined processes. By mimicking the energy-efficient behaviors observed in natural systems, these biomimetic approaches contribute to creating more sustainable and environmentally friendly solutions, aligning with the growing emphasis on energy conservation and eco-friendly technologies. In essence, these examples highlight the transformative potential of biomimicry in optimizing shapes, materials, repair mechanisms, and energy use, showcasing how nature's efficiency becomes a guiding principle for enhancing human-made products and technologies. (Verbrugghe et al., 2023).

Fig 5.3.1: Series of Solar PanelsSource: (*Free Images: Sky, Sun, Technology, Sunlight, Solar Panel, Solar Power, Solar Energy, Photovoltaics, Solar Panels 5184x3456 - - 708107 - Free Stock Photos - PxHere*, n.d.)

Adaptation to environmental changes is a fundamental trait observed in natural structures, and this principle serves as a valuable guide in biomimetic design. By incorporating adaptable features inspired by nature, biomimetic designs can enhance their performance in varying environmental conditions. Drawing inspiration from the way natural structures evolve and respond to changes, biomimetic designs showcase a dynamic and versatile approach to overcoming challenges.

Nature's proficiency in resource efficiency offers impactful lessons, such as the effective transport of water in trees. Biomimetic designs can leverage these insights to develop efficient water distribution systems, optimizing the use of resources in various applications. This application of biomimicry not only enhances efficiency but also contributes to sustainable practices, aligning with the increasing focus on responsible resource management.

Sustainability takes center stage in biomimetic design, with an emphasis on the use of sustainable and renewable materials. Biomimicry mirrors the cyclical and waste-minimizing processes observed in natural ecosystems, promoting the creation of materials that align with environmental conservation. This commitment to sustainable materials ensures that biomimetic designs contribute to reducing the ecological footprint, fostering a more harmonious relationship between human-made structures and the natural world. (E. Kennedy et al., 2015)

Integration with ecosystems emerges as a key principle influenced by lessons from natural environments. Biomimetic designs that draw insights from ecosystems aim to seamlessly integrate buildings and structures into their surroundings, promoting a sustainable coexistence with nature. This approach not only enhances the aesthetic appeal of structures but also contributes to the preservation of biodiversity and ecosystem health. In essence, these biomimetic principles of adaptation, resource efficiency, sustainable materials, and integration with ecosystems collectively illustrate how nature serves as an enduring source of inspiration for creating designs that harmonize with the environment and contribute to a more sustainable future. (Perera & Coppens, n.d.; Taieb & Amor, n.d.)

Fig. 5.3.2: Building Covered by Green PlantsSource:https://www.peakpx.com/407097/gray-high-rise-building-with-green-plants

These lessons from nature contribute to the development of more efficient, sustainable, and environmentally friendly structures in various industries, from architecture to transportation and beyond. Biomimetic structures not only reduce environmental impact but also often lead to more cost-effective and long-lasting solutions.

Summary

This chapter delves into the fascinating realm of biomimetic structures and design, showcasing how insights from nature inspire innovative solutions across diverse fields. It begins by exploring biophilic design, which seamlessly integrates natural elements into architectural spaces to enhance human well-being. Drawing inspiration from termite mound architecture, biomimetic buildings employ passive cooling and heating systems, reducing energy consumption and environmental impact. Furthermore, biomorphic architecture incorporates organic shapes and patterns, exemplified by structures like the Eden Project's biomes, which prioritize both aesthetics and sustainability. The discussion extends to bionics, where biomimetic principles inform the design of lightweight structures, as seen in the Water Cube in Beijing, inspired by soap bubbles. Green roofs and living walls exemplify biophilic design strategies that promote sustainability and improve urban environments. Additionally, the text explores the use of sustainable materials like bamboo and the optimization of solar energy, demonstrating how biomimetic approaches revolutionize architectural design while fostering environmental stewardship. Transitioning to aerospace and vehicle design, the text illustrates how biomimicry drives innovation in these industries. Engineers draw inspiration from bird wings to enhance aircraft performance, incorporating winglets for improved aerodynamics and quieter designs inspired by owl wings. Biomimetic drones and vehicles emulate the agility of natural flyers, enabling precise maneuvers for applications such as surveillance and exploration. Lightweight and robust materials inspired by biological structures, such as bones, contribute to fuel efficiency and structural integrity in aerospace technologies. Renewable energy applications benefit from biomimetic wind turbine blade designs, influenced by humpback whale flippers, to optimize energy generation.

References

A bird flying in the sky with its wings spread. Animal bird fly, animals. - PICRYL - Public Domain Media Search Engine Public Domain Image. (n.d.). Retrieved May 9, 2024, from https://garystockbridge617.getarchive.net/amp/media/animal-bird-fly-animals-aead5f

Araque, K., Palacios, P., Mora, D., & Austin, M. C. (2021). Biomimicry-Based Strategies for Urban Heat Island Mitigation: A Numerical Case Study under Tropical Climate.

Bilici, S. C., Küpeli, M. A., & Guzey, S. S. (n.d.-a). Inspired by nature: An engineering design-based biomimicry activity.

Fiorentino, C., & Montana-Hoyos, C. (n.d.). The Emerging Discipline of Biomimicry as a Paradigm Shift towards Design for Resilience.

Free Images: Sky, sun, technology, sunlight, solar panel, solar power, solar energy, photovoltaics, solar panels 5184x3456—- 708107—Free stock photos—PxHere. (n.d.). Retrieved May 9, 2024, from https://pxhere.com/en/photo/708107

Huh, D., Torisawa, Y., Hamilton, G. A., Kim, H. J., & Ingber, D. E. (2012). Microengineered physiological biomimicry: Organs-on-chips. Lab on a Chip, 12(12), 2156–2164. https://pubs.rsc.org/en/content/articlehtml/2012/lc/c2lc40089h

Ilieva, L., Ursano, I., Traista, L., Hoffmann, B., & Dahy, H. (2022). Biomimicry as a Sustainable Design Methodology—Introducing the 'Biomimicry for Sustainability' Framework.

Kennedy, E. B., & Marting, T. A. (2016). Biomimicry: Streamlining the Front End of Innovation for Environmentally Sustainable Products: Biomimicry can be a powerful design tool to support sustainability-driven product development in the front end of innovation. Research-Technology Management, 59(4), 40–48. https://doi.org/10.1080/08956308.2016.1185342

Kennedy, E., Fecheyr-Lippens, D., Hsiung, B.-K., Niewiarowski, P. H., & Kolodziej, M. (2015). Biomimicry: A Path to Sustainable Innovation. 31(3).

Perera, A. S., & Coppens, M.-O. (n.d.). Re-designing materials for biomedical applications: From biomimicry to nature-inspired chemical engineering.

Taieb, A. H., & Amor, M. B. (n.d.). Eco-design Design Strategy: Case Study Biomimicry Approach in Design Product Materials. 21(3).

Verbrugghe, N., Rubinacci, E., & Khan, A. Z. (2023). Biomimicry in Architecture: A Review of Definitions, Case Studies, and Design Methods.

CHAPTER VI

Bio-Inspired Sensors and Sensing

This chapter is basically based on Bio-Inspired Sensors and Sensing summarizes the exploration of nature-inspired sensor technologies and their applications across various fields. It delves into the mechanisms found in nature, such as echolocation in bats, antennae sensing in insects, and human sensory systems, which serve as models for innovative sensor design. The chapter highlights how bio-inspired sensors have revolutionized environmental monitoring, healthcare, robotics, aerospace, agriculture, and more. It discusses specific applications in engineering challenges, such as structural health monitoring, aircraft design, underwater robotics, materials research, energy harvesting, and noise reduction. Additionally, it explores the role of bio-inspired sensors in fields like geotechnical engineering, fire detection, and space exploration, showcasing their versatility and potential for advancing technology and addressing complex engineering problems.

6.1 Sensing Mechanisms in Nature

Bio-inspired sensors and sensing mechanisms draw inspiration from nature to develop innovative sensor technologies. Nature has evolved remarkable sensing mechanisms in various organisms, which can serve as models for creating advanced sensors. Some examples of sensing mechanisms in nature include:

Nature's intricate designs continue to inspire innovations in various fields, particularly in the realm of sensory technology. Take, for instance, the remarkable echolocation ability observed in bats. By emitting high-frequency sound waves and interpreting the returning echoes, bats adeptly navigate and locate prey. This natural phenomenon serves as the foundation for the development of sonar and ultrasonic sensors, finding applications in diverse fields such as robotics and medical imaging. The principles derived from the sophistication of bat echolocation contribute to the creation of advanced sensor technologies that enhance our ability to perceive and understand the environment.(Ronkainen et al., 2010)

Similarly, the world of insects offers another wellspring of inspiration. Insects like moths and bees possess highly sensitive antennae, allowing them to detect

subtle chemical cues, including pheromones. This extraordinary capability has become a muse for scientists and engineers working on chemical sensors. These sensors, influenced by insect antennae, find utility in detecting gases, odors, and pollutants. The marriage of biological insights from insects with technological advancements has paved the way for the development of sophisticated sensors, mimicking nature's precision and sensitivity in detecting chemical signals.(Turner, 2013)

Fig. 6.1 A beetle on a leaf(Morin, 2018)

The marvels of human biology provide a rich source of inspiration for cutting-edge technologies, particularly in the domain of sensory systems. Human vision, a complex interplay of optics and neural processing, has given rise to the development of cameras and imaging sensors designed to replicate the structure and functionality of the human eye. This biomimetic approach not only enhances our understanding of visual perception but also contributes to advancements in imaging technologies used across various industries.(Ronkainen et al., 2010)

Moving beyond sight, the human skin's remarkable ability to sense pressure, temperature, and texture has spurred the creation of artificial skin and tactile sensors. These sensors, mirroring the tactile capabilities of human skin, find applications in robotics and prosthetics, where the sense of touch is crucial for tasks ranging from delicate manipulation to ensuring user safety.(Taieb & Amor, n.d.)

Nature's ingenuity extends to avian navigation, where birds like pigeons utilize Earth's magnetic field for orientation. This phenomenon has inspired the development of magnetic field sensors and compasses, showcasing how biomimicry can yield innovative solutions for navigation technologies.

Delving into the depths of the ocean, the bioluminescence displayed by deep-sea creatures for purposes like camouflage and communication has influenced the creation of light-sensitive sensors and imaging technologies. These technologies prove invaluable in scenarios with minimal light conditions, offering insights into the diverse adaptations found in nature.(Turner, 2013)

In essence, the emulation of human and animal sensory systems through biomimicry not only pushes the boundaries of technological innovation but also underscores the efficiency and sophistication inherent in natural designs.(Bilici et al., n.d.)

The intricate mechanisms of thermoregulation observed in reptiles, exemplified by the desert iguana's use of specialized scales to manage body temperature through sunlight absorption and reflection, have spurred innovative applications in material science and architecture. Drawing inspiration from nature's thermal regulation, biomimicry has led to the development of smart materials that mimic these reptilian adaptations. These materials are designed for temperature regulation, demonstrating the potential for energy-efficient solutions in various industries. Additionally, the principles derived from reptilian thermoregulation have been incorporated into the design of buildings, contributing to the creation of structures that optimize energy consumption through efficient temperature control. This biomimetic approach not only harnesses the efficiency of natural adaptations but also provides sustainable solutions for enhancing thermal comfort in diverse environments.(McConney et al., 2009; Perera & Coppens, n.d.; Taieb & Amor, n.d.)

These natural sensing mechanisms provide valuable insights for scientists and engineers in designing sensors with enhanced sensitivity, efficiency, and adaptability. By mimicking nature, bio-inspired sensors can offer innovative solutions in a wide range of applications, from healthcare and environmental monitoring to aerospace and robotics.(McConney et al.,

2009; Perera & Coppens, n.d.)

6.2 Application in Sensor Technology

Bio-inspired sensors and sensing techniques have found numerous applications in sensor technology across various fields. These applications leverage the efficiency and adaptability of natural sensing mechanisms to create innovative and advanced sensor systems. Here are some key applications of bio-inspired sensors in sensor technology:

Biomimicry finds practical applications across diverse sectors, showcasing its transformative impact on technology and innovation. In the realm of environmental monitoring, biomimetic chemical sensors stand out for their emulation of animal and insect olfactory systems. These sensors prove instrumental in detecting environmental pollutants, gases, and toxins, contributing to air quality monitoring, industrial safety, and early warning systems for chemical leaks. By replicating nature's keen sense of smell, biomimetic chemical sensors enhance our ability to assess and mitigate environmental risks.(Perera & Coppens, n.d.; Ronkainen et al., 2010)

The field of medical and healthcare technology benefits significantly from biomimicry, particularly through biosensors inspired by the human body's biomarker detection capabilities. Used in diagnosing diseases, monitoring glucose levels, and detecting pathogens, biosensors play a vital role in medical devices like glucose meters and lab-on-a-chip systems. This bio-inspired approach not only advances diagnostic capabilities but also streamlines healthcare processes.

In robotics and automation, the integration of bio-inspired tactile sensors marks a notable development. These sensors emulate the human sense of touch, providing robots with the ability to sense and manipulate objects with sensitivity and dexterity. This innovation contributes to the advancement of robotic systems, making them more versatile and adaptable to various tasks.(Perera & Coppens, n.d.)

The aerospace and navigation industry benefits from biomimetic principles, particularly in bio-inspired navigation systems. Sensors inspired by animals like birds, which utilize magnetic fields for navigation, find applications in aerospace and marine navigation systems. These sensors can be implemented in compasses and GPS technology, enhancing the precision and efficiency of navigation systems.(Perera & Coppens, n.d.; Turner, 2013)

Biomimicry extends its influence to imaging and vision systems through biologically-inspired cameras. These cameras are designed based on the human eye's structure and visual processing, significantly impacting photography, surveillance, medical imaging, and autonomous vehicles. By replicating the sophistication of the human eye, biologically-inspired cameras contribute to better image capture and processing, advancing the capabilities of various visual technologies. In essence, these real-world applications highlight the versatility and ingenuity of biomimicry, showcasing its potential to revolutionize technology across multiple industries.(Amer, 2019; Perera & Coppens, n.d.)

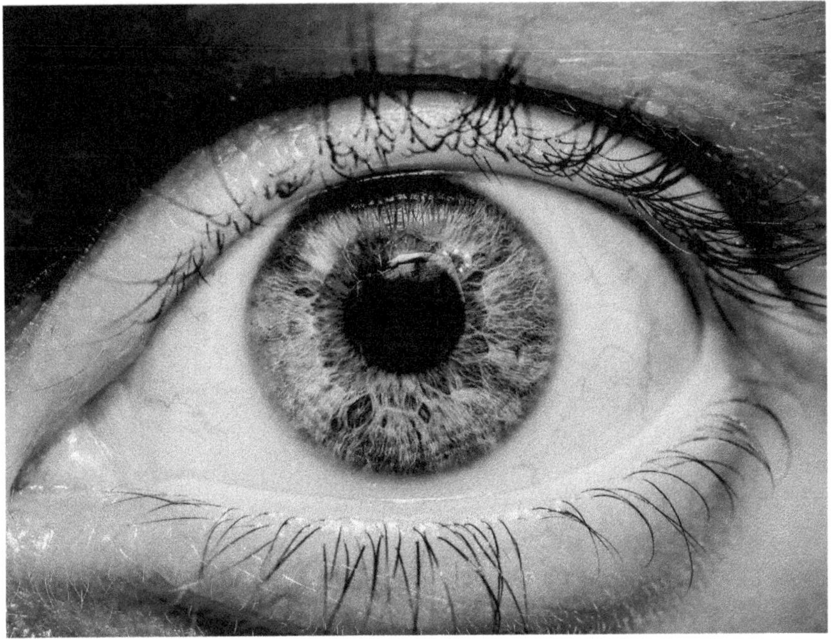

Fig. 6.2: Human Eye

Biomimicry manifests its influence in various domains, further exemplified by applications in underwater exploration, energy efficiency , and agriculture.

In the realm of underwater exploration, bioluminescence sensors stand as a testament to biomimicry's contribution. Inspired by bioluminescent organisms, these sensors are instrumental in underwater exploration and deep-sea research. They possess the capability to detect and analyze bioluminescent phenomena in the ocean, offering invaluable insights into the mysterious and often uncharted realms of the underwater world.(Amer, 2019; Kennedy et al., 2015)

The integration of biomimetic principles into energy-efficient architecture is evident through thermoregulation sensors. These sensors draw inspiration from certain animals' adeptness at regulating their body temperature by either reflecting or absorbing sunlight. Applied in building design, these sensors contribute to energy-efficient climate control, optimizing the use of natural resources and minimizing environmental impact.(Amer, 2019; Kennedy et al., 2015)

This biomimetic approach not only enhances sustainability in architecture but also highlights the potential for eco-friendly solutions inspired by nature.

Agriculture benefits from biomimicry in the form of chemical sensing inspired by insects. Sensors designed to mimic insects' remarkable ability to detect chemical cues find applications in agriculture, particularly in monitoring pests and implementing targeted pesticide control. This bio-inspired approach contributes to more precise and environmentally conscious pest management strategies, aligning with the growing emphasis on sustainable agricultural practices.(Biagioni & Bridges, n.d.; Blok & Gremmen, 2016)

These biomimetic applications in underwater exploration, energy-efficient architecture, and agriculture underscore the versatility of nature-inspired solutions. By drawing inspiration from the intricacies of the natural world, biomimicry continues to offer innovative and sustainable approaches across diverse fields, pushing the boundaries of human ingenuity.s(Amer, 2019; Bilici et al., n.d.; Ronkainen et al., 2010)

Fig. 6.3: A Red Beetle(Smyczek, 2019)

Biomimicry plays a crucial role in enhancing search and rescue operations and expanding the capabilities of space exploration. In the realm of search and rescue operations, biomimetic sensors designed to replicate the keen senses of search and rescue animals, particularly dogs, have become invaluable tools. These sensors are employed in disaster-stricken areas to locate survivors and detect hazardous materials efficiently. By emulating the acute sensory abilities of animals, biomimetic sensors contribute to the effectiveness and precision of search and rescue efforts, ultimately improving the chances of saving lives and mitigating the impact of disasters.(Perera & Coppens, n.d.; Turner, 2013)

The application of biomimicry extends its reach to space exploration, where bio-inspired sensors hold promise for planetary exploration. Designed to mimic natural sensing mechanisms, these sensors can be integrated into planetary rovers, aiding in the detection of signs of life and analysis of environmental conditions on other celestial bodies. By drawing inspiration from the adaptability and sensing capabilities observed in living organisms, biomimetic sensors contribute to the advancement of space exploration, enhancing our understanding of distant planets and the potential for extraterrestrial life. In essence, these biomimetic applications in search and rescue operations and space exploration exemplify the versatility of nature-inspired solutions in addressing complex challenges, whether on Earth or in the vastness of outer space.(Perera & Coppens, n.d.; Ronkainen et al., 2010; Turner, 2013)

Bio-inspired sensors and sensing mechanisms offer innovative solutions that enhance the sensitivity, accuracy, and adaptability of sensor technology. By drawing inspiration from nature, researchers and engineers continue to develop cutting-edge sensor systems with a wide range of practical applications.

6.3 Sensing Solutions for Engineering Challenges

Bio-inspired sensors and testing methods have been instrumental in addressing various engineering challenges by offering innovative solutions inspired by nature. Here are some examples of sensing solutions for engineering challenges: In the field of structural engineering, bio-inspired sensors are making strides in Structural Health Monitoring (SHM). Taking inspiration from the adaptive mechanisms observed in natural structures like trees and bones, these sensors are deployed to monitor the health and integrity of various man-made structures, including bridges, buildings, and aircraft. Drawing parallels with biological systems, these sensors can detect changes in strain, stress, and vibrations, providing invaluable insights into the structural conditions. This bio-inspired approach to SHM is instrumental in preventing failures, ensuring early detection of potential issues, and ultimately contributing to enhanced safety in infrastructure.(McConney et al., 2009)

Aircraft design has also embraced bio-inspiration, specifically in the improvement of wing structures. Sensors inspired by the wings of birds have been incorporated to enhance both the aerodynamic performance and structural integrity of aircraft wings. By emulating the flexible and adaptable nature of avian wings, engineers aim to create aircraft with more efficient and safer wing designs. This bio-inspired integration not only contributes to advancements in aviation technology but also emphasizes the potential of nature-inspired solutions in optimizing the performance and safety of man-made structure.(Bérut & Grossman, 2011) (Hwang et al., n.d.)

In the domain of underwater robotics, bio-inspired sensors play a pivotal role, taking inspiration from fish lateral line systems. These sensors are employed in underwater robots to emulate the remarkable ability of fish to detect obstacles and currents. By replicating this natural sensing mechanism, underwater robots equipped with these sensors can navigate more efficiently and maneuver adeptly in challenging underwater environments. This bio-inspired approach not only enhances the operational capabilities of underwater robotics but also underscores the potential for nature-inspired solutions in advancing robotics for specific applications.(Amer, 2019; Kennedy et al., 2015)

In the realm of materials research, bio-inspired sensors find application, particularly those inspired by the color-changing ability of animals like chameleons. These sensors contribute to the study and development of smart materials that can dynamically alter their properties, such as color, in response to environmental conditions. This bio-inspired approach to materials research opens avenues for the creation of adaptive and responsive materials, with potential applications in various fields, from design to engineering Bio-inspired sensors are also making an impact in the field of energy harvesting. Drawing inspiration from the way plants capture and convert solar energy, these sensors are integrated into energy harvesting technologies such as solar panels and piezoelectric devices. By replicating nature's efficient energy conversion processes, these sensors contribute to improving the overall energy efficiency and sustainability of harvesting technologies.("Biomimicry, an Approach, for Energy Effecient Building Skin Design," 2016; "Ecological Sustainability, Nature –Inspired, Architecture, Zero- Waste Systems, Regenerative Design," 2018)

In the pursuit of noise reduction technology, bio-inspired sensors inspired by the silent hunting capabilities of owls have been applied. These sensors, mimicking the owl's ability to hunt silently, are utilized to detect and mitigate unwanted sounds in various engineering applications. This bio-inspired approach to noise reduction demonstrates the versatility of nature-inspired solutions, not only in robotics and materials but also in addressing challenges related to sound and acoustics in engineering contexts.

Fig. 6.4: Birds Feather(Head, 2018)

Bio-inspired sensors take cues from fish scales and the movements of aquatic creatures. These sensors are strategically employed to optimize fluid flow control in various industrial processes, ranging from pipelines to HVAC systems. By mimicking nature's efficient mechanisms for navigating through fluid environments, these bio-inspired sensors contribute to enhancing the efficiency and performance of fluid-related systems, providing a promising avenue for innovation in industrial applications.(Ronkainen et al., 2010; Turner, 2013)

Bio-inspired sensors find significant application in the development of adaptive robotic limbs and exoskeletons. Drawing inspiration from the intricate sensing mechanisms observed in nature, these sensors are integrated into robotic systems to enhance their adaptability, responsiveness, and safety. This bio-inspired approach extends its utility across diverse domains, from aiding in medical rehabilitation to improving efficiency in industrial assembly tasks. The incorporation of nature-inspired sensors in robotic technology exemplifies the potential for bio-inspired solutions to revolutionize fields that require precision, dexterity, and safety.("Biomimicry, an Approach, for Energy Effecient Building Skin Design," 2016; Turner, 2013)

In the agricultural domain, bio-inspired sensors play a crucial role in precision farming. Taking inspiration from the sensing mechanisms found in plants and insects, these sensors are utilized for soil moisture monitoring, pest detection, and overall crop management. By replicating the natural processes that contribute to plant health and protection, bio-inspired sensors contribute to sustainable and efficient practices in modern agriculture.(Blok & Gremmen, 2016)

Bio-inspired sensors extend their influence to geotechnical engineering, where they are employed for soil and ground condition monitoring. Inspired by burrowing animals like moles, these sensors assist engineers in better understanding and responding to soil movement and stability. This bio-inspired approach enhances the capabilities of geotechnical engineering, providing valuable insights into subsurface conditions and facilitating more informed decision-making in construction and infrastructure projects.("Biomimicry, an Approach, for Energy Effecient Building Skin Design," 2016; Turner, 2013)

In fire detection systems, bio-inspired sensors draw inspiration from animals with an acute sense of smell, such as dogs. Applied to fire detection and suppression, these sensors exhibit high sensitivity to smoke and fire-related gases. This bio-inspired technology enhances the early detection of fire hazards, contributing to improved safety measures and more effective fire prevention strategies.

Fig6.5: Fire detection alarm system(Schumin, 2013)

Bio -inspired sensors emerge as valuable tools for enhancing the capabilities of telescopes and space instruments. Drawing inspiration from the intricate sensing mechanisms found in living organisms, these bio-inspired sensors contribute to the improvement of detection and analysis processes related to cosmic phenomena. Whether replicating the precision of biological sensory systems or adapting principles from nature to space-related challenges, bio-inspired sensors offer innovative solutions that can elevate the performance and functionality of instruments used in the exploration of the cosmos. This application of bio-inspired technology demonstrates the versatility and adaptability of nature-inspired approaches, showcasing their potential to drive advancements not only in engineering and industry but also in the broader realms of scientific discovery and space exploration. Through the integration of bio-inspired sensors, the field of astronomy and space exploration stands to benefit from enhanced sensitivity, adaptability, and overall efficiency in observing and understanding the mysteries of the universe.("Biomimicry, an Approach, for Energy Effecient Building Skin Design," 2016; "Biomimicry, an

Approach, for Energy Effecient Building Skin Design," 2016; "Ecological Sustainability, Nature –Inspired, Architecture, Zero- Waste Systems, Regenerative Design," 2018; Turner, 2013).

Summary

A summary of this chapter based on Bio-Inspired Sensors and Sensing provides a condensed overview of the exploration into nature-inspired sensor technologies and their diverse applications. It outlines the various mechanisms observed in nature, such as echolocation in bats, antennae sensing in insects, and human sensory systems, which serve as inspiration for innovative sensor design. The summary highlights the wide-ranging applications of bio-inspired sensors in fields like environmental monitoring, healthcare, robotics, aerospace, and agriculture. It also discusses specific examples of how these sensors address engineering challenges, such as structural health monitoring, aircraft design, underwater robotics, and materials research. Overall, the summary encapsulates the chapter's discussion on the versatility and potential of bio-inspired sensors to drive technological advancements and solve complex engineering problems.

References

Amer, N. (2019). Biomimetic Approach in Architectural Education: Case study of 'Biomimicry in Architecture' Course. *Ain Shams Engineering Journal.*

Bérut, O., & Grossman, S. (2011). Innover, c'est l'affaire de tous. *tech N o log i e.*

Biagioni, E. S., & Bridges, K. W. (n.d.). *THE APPLICATION OF REMOTE SENSOR TECHNOLOGY TO ASSIST THE RECOVERY OF RARE AND ENDANGERED SPECIES.*

Bilici, S. C., Küpeli, M. A., & Guzey, S. S. (n.d.). *Inspired by nature: An engineering design-based biomimicry activity.*

Biomimicry, an Approach, for Energy Effecient Building Skin Design. (2016). *Procedia Environmental Sciences.*

Blok, V., & Gremmen, B. (2016). Ecological Innovation: Biomimicry as a New Way of Thinking and Acting Ecologically. *Journal of Agricultural and Environmental Ethics, 29*(2), 203–217. https://doi.org/10.1007/s10806-015-9596-1

Ecological sustainability, Nature –inspired, Architecture, Zero- waste systems, Regenerative design. (2018). *Architecture Research.*

Kennedy, E., Fecheyr-Lippens, D., Hsiung, B.-K., Niewiarowski, P. H., & Kolodziej, M. (2015). *Biomimicry: A Path to Sustainable Innovation. 31*(3).

McConney, M. E., Anderson, K. D., Brott, L. L., Naik, R. R., & Tsukruk, V. V. (2009).

Bioinspired Material Approaches to Sensing. *Adv. Funct. Mater.*

Morin, B. (2018). *English: Xylotrupes socrates (Siamese rhinoceros beetle, or "fighting beetle"), male, on a banana leaf, Focus stacking from 23 images. This scarab beetle is particularly known for its role in insect fighting in Northern Laos and Thailand. Own work.*

https://commons.wikimedia.org/wiki/
File:Xylotrupes_socrates_%28Siamese_rhinoceros_beetle%29.jpg

Perera, A. S., & Coppens, M.-O. (n.d.). *Re-designing materials for biomedical applications: From biomimicry to nature-inspired chemical engineering.*

Ronkainen, N. J., Halsall, H. B., & Heineman, W. R. (2010). *Electrochemical biosensors.*

Smyczek, M. (2019). *English: This is a kind of beetle, known as the scalet lily beetle. But it is different because the head isn't black - and it's red. Own work.*

https://commons.wikimedia.org/wiki/
File:Beetle_on_a_green_plant.jpg

Taieb, A. H., & Amor, M. B. (n.d.). *Eco-design Design Strategy: Case Study Biomimicry Approach in Design Product Materials. 21*(3).

Turner, A. P. F. (2013). Biosensors: Sense and sensibility. *Chem Soc Rev.*

CHAPTER VII

Bio-Inspired Robotics and Automation

This chapter is basically Bio-Inspired Robotics and Automation provides a concise overview of the chapter's contents and key themes. It outlines how biomimicry serves as a transformative force in various fields, including search and rescue operations, space exploration, industrial automation, and environmental monitoring. The abstract highlights the role of biomimetic sensors in enhancing search and rescue efforts and planetary exploration, as well as the design and functionality of biomimetic robots in tasks such as climbing, gripping, and collaborative work. Additionally, it previews future trends in bio-inspired robotics, including soft robotics, biohybrid robots, and the integration of artificial intelligence. Overall, the abstract encapsulates the chapter's discussion on the applications and potential of biomimetic robotics in addressing complex challenges across industries.

7.1 Robotics Inspired by Natural Movements

Biomimicry serves as a transformative force in both search and rescue operations and the realm of space exploration, offering innovative solutions inspired by the natural world. In the critical domain of search and rescue operations, biomimetic sensors have been designed to replicate the acute senses of search and rescue animals, particularly drawing inspiration from the capabilities of dogs. These sensors are deployed in disaster-stricken areas to enhance the efficiency of locating survivors and detecting hazardous materials. By emulating the remarkable sensory acuity of animals, biomimetic sensors bring an unprecedented level of precision to search and rescue efforts, significantly improving the effectiveness of operations in challenging and dynamic environments. This application of biomimicry not only showcases the potential for technology to mirror nature's capabilities but also underscores its humanitarian impact in crisis situations.(Turner, 2013)

The influence of biomimicry extends to the domain of space exploration, where bio-inspired sensors open new frontiers for planetary exploration. Integrated into planetary rovers, these sensors replicate natural sensingmechanisms, enabling the detection of signs of life and the analysis

67

of environmental conditions on celestial bodies beyond Earth. By harnessing principles observed in living organisms, biomimetic sensors contribute to the advancement of space exploration, providing valuable insights into the potential habitability of other planets and the search for extraterrestrial life. This dual application of biomimicry highlights its versatility in addressing challenges across diverse environments, from disaster-stricken regions on Earth to the uncharted territories of our solar system.(*Steampunk Submarine by LG-Design on DeviantArt*, n.d.)

Fig7.1: A model depicting submarine inspired from fish(Turner, 2013)

Biomimicry finds innovative applications in the realm of robotics, influencing the design and functionality of crawling, climbing robots, robotic hands, grippers, and collaborative robots (cobots).

In the domain of crawling and climbing robots, gecko-inspired climbers showcase the emulation of nature's adhesive mechanisms found in gecko feet. These robots employ adhesive materials to climb walls and ceilings, enabling inspection and maintenance tasks in challenging environments where conventional methods may be impractical. Additionally, insect-inspired crawlers mimic the movement patterns of ants and cockroaches, making them ideal for search and rescue missions in disaster scenarios. By replicating the agility and adaptability of these natural organisms, biomimetic robots enhance their effectiveness in navigating complex and unpredictable environments.(Araque et al., 2021; Coyle et al., n.d.)

Robotic hands and grippers benefit from biomimicry through the creation of anthropomorphic hands that closely replicate the dexterity and grasping capabilities of the human hand. These hands find applications in tasks such as assembly and surgery, where precise and flexible manipulation is crucial. Furthermore, soft grippers draw inspiration from the flexibility and adaptability observed in natural organisms, making them well-suited for handling delicate objects and navigating uncertain environments. This biomimetic approach enhances the versatility and safety of robotic interactions in various industries.(Neveln et al., n.d.)

In the realm of collaborative robots, or cobots, biomimicry takes inspiration from social insects like ants and bees. These robots are designed based on the teamwork and coordination observed in social insect colonies. Applied in tasks such as warehouse automation and logistics, cobots showcase the efficiency and collaborative capabilities derived from natural systems. This integration of biomimicry into robotic technologies not only enhances their performance but also opens new possibilities for human-robot collaboration in diverse applications, ultimately advancing the field of robotics in alignment with principles observed in the natural world.(Coyle et al., n.d.; Neveln et al., n.d.)

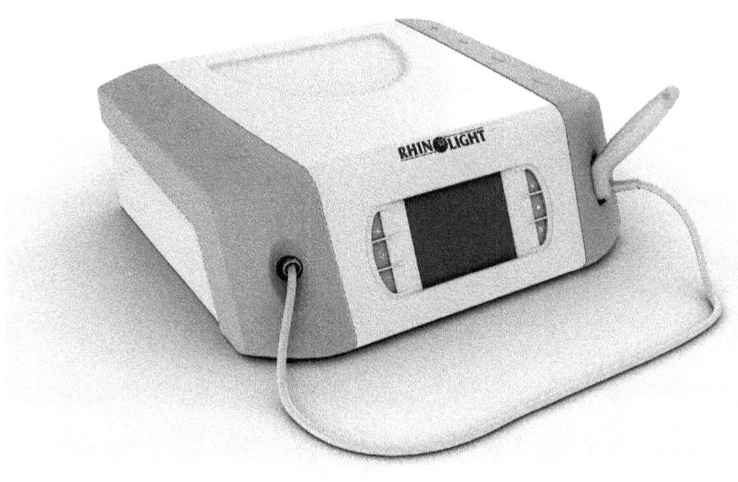

Fig 7.2: An EMF machine

In the realm of robotics, biomimicry extends its influence to innovative sensors and perception as well as bio-inspired behavior and learning.

Bio-inspired sensors represent a significant advancement, emulating the sensing mechanisms found in the animal kingdom. Vision systems inspired by the human eye and sonar systems inspired by bats are notable examples. These sensors enhance a robot's perception, enabling it to navigate and interact with its environment more effectively. By drawing inspiration from the intricate sensing capabilities of living organisms, these bio-inspired sensors contribute to the adaptability and responsiveness of robots, bridging the gap between artificial intelligence and natural perception.(Araque et al., 2021; Turner, 2013)

Bio-inspired behavior and learning in robotics take shape through neuromorphic control. This approach involves leveraging biologically-inspired neural networks to govern robot behavior and learning processes. By mimicking the neural structures observed in living organisms, robots equipped with neuromorphic control can adapt and self-improve over time. This bio-inspired learning capability allows robots to continuously refine their actions and responses based on experience, leading to more intelligent and flexible robotic systems. The integration of biomimicry into sensors, perception, and learning mechanisms not only enhances the capabilities of robots but also aligns artificial intelligence more closely with the sophisticated processes observed in the natural world.(Neveln et al., n.d.; Turner, 2013)

By imitating natural movements and behaviors, bio-inspired robotics and automation systems can excel in a wide range of applications, from industrial automation to disaster response, space exploration, and medical surgery. These robots are designed to be more energy-efficient adaptable to various environments, and capable of performing tasks that would be challenging for traditional robots.

7.2 Automation and Industry Applications

Bio-inspired robotics and automation have made significant contributions to various automation and industrial applications, offering solutions that enhance efficiency, adaptability, and productivity. Here are some notable applications in the industrial sector:

The integration of biomimicry in robotics has revolutionized various industries, showcasing its impact in manufacturing and assembly, warehouse logistics, and agriculture. In the realm of manufacturing and assembly, robots drawing inspiration from the dexterity of the human hand have found a pivotal role in assembly lines.

These robots excel in performing intricate tasks, such as precision assembly and quality control, contributing to increased efficiency and accuracy in manufacturing processes. Collaborative robots, or cobots, represent another facet of biomimicry in manufacturing. Working alongside human workers, cobots enhance productivity and safety by combining the strengths of human intuition and robot precision, fostering a collaborative and efficient manufacturing environment.(Neveln et al., n.d., n.d.)

Warehouse and logistics benefit significantly from biomimetic applications, particularly in the form of autonomous robots inspired by insect swarming behavior. Emulating the efficiency and coordination observed in swarming insects, these robots navigate warehouses with autonomy, managing inventory and optimizing order fulfillment. This biomimetic approach not only streamlines logistical operations but also reduces operational costs through enhanced automation and organization.(Coyle et al., n.d.; Neveln et al., n.d.)

In the agricultural sector, bio-inspired robots equipped with advanced vision systems and machine learning algorithms have ushered in a new era of precision agriculture. Taking cues from the human eye's perception or employing learning mechanisms inspired by nature, these robots automate the monitoring and management of crops and livestock. The result is a more efficient and sustainable agricultural practice that maximizes yield while minimizing resource usage. This integration of biomimicry in manufacturing, logistics, and agriculture exemplifies the transformative potential of nature-inspired solutions in enhancing productivity, efficiency , and sustainability across diverse industries.(Blok & Gremmen, 2016; Coyle et al., n.d.)

Fig7.3: A building covered with green plantshttps://www.peakpx.com/407097/gray-high-rise-building-with-green-plants

Biomimicry has left an indelible mark on industries ranging from mining and construction to pharmaceuticals and food processing, introducing transformative applications in various sectors.

In mining and construction, robotic vehicles inspired by the agility of insects or reptiles navigate remote and hazardous locations, undertaking tasks such as exploration, mapping, and inspections. This biomimetic approach enhances efficiency and safety in industries where traditional methods might be impractical or risky.(Neveln et al., n.d.)

For energy generation and infrastructure maintenance, climbing robots, drawing inspiration from geckos, inspect and upkeep critical structures like bridges, dams, and wind turbines. By reducing the reliance on human interventions in precarious environments, these robots enhance the safety and precision of maintenance procedures.("Biomimicry, an Approach, for Energy Effecient Building Skin Design," 2016)

In pharmaceuticals and the chemical industry, robots mimicking the movement of biological organisms handle hazardous materials, automate laboratory processes, and contribute to drug discovery. This application of biomimicry improves precision and safety in industries where accuracy is paramount.(Dumanli & Savin, 2016; Perera & Coppens, n.d.)

The food processing and packaging industry benefit from robots equipped with soft grippers and dexterous hands inspired by natural organisms. These robots excel in delicate tasks like picking, sorting, and packaging of food items, ensuring efficiency and maintaining product integrity.(Ruiz-Garcia, 2009)

Quality control and inspection in manufacturing processes integrate vision systems inspired by animal vision. These systems, mimicking the sophistication of natural visual processes, are deployed on production lines to inspect and detect defects in products, ensuring high-quality output.

Waste management and recycling witness the utilization of robots inspired by the foraging behavior of animals. In waste sorting and recycling facilities, these robots enhance efficiency and contribute to reducing the environmental impact of waste disposal.

Biomimetic 3D printers have revolutionized additive manufacturing by drawing inspiration from the structure and construction techniques found in nature. This innovation allows for more efficient and adaptable 3D printing processes, shaping the future of manufacturing.(Neveln et al., n.d.)

In material handling and conveyance, conveyor systems inspired by the cooperative behavior of ants or other social insects optimize the flow of materials in industrial settings. By replicating the efficiency observed in natural systems, these biomimetic conveyors contribute to streamlined industrial operations. Overall, these examples underscore the diverse and impactful applications of biomimicry across a spectrum of industries, showcasing its potential to revolutionize processes, enhance safety, and promote sustainability.(Dumanli & Savin, 2016; Perera & Coppens, n.d.)

In the realm of environmental monitoring and cleanup, biomimicry finds compelling applications through autonomous robots inspired by marine life or birds. These robots play a crucial role in monitoring natural and industrial environments, conducting tasks such as environmental assessments, water quality evaluations, and pollution cleanup. Drawing inspiration from the intricate movements and adaptive behaviors observed in marine organisms and birds, these autonomous robots navigate diverse and challenging terrains, contributing to the preservation of ecosystems and the mitigation of environmental damage(Bensaude-Vincent, n.d.; Hayes et al., 2019).

The integration of bio-inspired robotics and automation not only enhances the efficiency and productivity of industrial processes but also fosters safety and sustainability. By deploying robots in environments where human intervention might be hazardous or impractical, these applications exemplify the potential of biomimetic design principles to address complex challenges in industry. The use of nature-inspired technologies in environmental monitoring and cleanup showcases the versatility of biomimicry, not only as a tool for innovation and efficiency but also as a means to promote environmental stewardship and sustainable practices in industrial settings.(Coyle et al., n.d.; Neveln et al., n.d.)

7.3 Future Trends in Bio-Inspired Robotics

Future trends in bio-inspired robotics and automation are poised to bring about significant advancements in various industries. As technology and research continue to progress, here are some of the key trends we can expect to see in bio-inspired robotics: The future of biomimetic robotics holds exciting developments across various domains. Soft robotics, drawing inspiration from the flexibility of natural organisms, is poised for significant growth. These robots, characterized by their pliable and adaptable structures, are well-suited for navigating unstructured environments, interacting safely with humans, and offering innovative solutions in fields such as medical surgery and search and rescue operations.(Coyle et al., n.d.; Neveln et al., n.d.)

Biohybrid robots represent a promising trend, integrating biological components with artificial systems. By incorporating living cells, tissues, or organisms, these robots aim to enhance adaptability and efficiency. Applications in environmental monitoring and healthcare stand to benefit from the synergistic integration of biological and artificial elements.(Neveln et al., n.d., n.d.)

The evolution of neuromorphic control is set to play a pivotal role in the advancement of biomimetic robotics. Bio-inspired neural networks and neuromorphic computing will empower robots to learn, adapt, and make real-time decisions, closely mirroring the cognitive processes observed in living organisms.

Swarm robotics, inspired by the collective behavior of social insects and animals, will continue to progress. The collaborative efforts of small robots working together hold promise for accomplishing intricate tasks, including environmental monitoring, exploration, and disaster response. This collaborative and bio-inspired approach to robotics opens new horizons for creating intelligent, adaptive, and efficient robotic systems with applications spanning a diverse range of industries.(Coyle et al., n.d.; Hayes et al., 2019)

The future trajectory of biomimetic robotics promises transformative advancements across various applications. Birds and insects are poised to play an increasingly common role in surveillance, environmental monitoring, and delivery services, leveraging their natural capabilities for efficient and agile navigation.

Human-robot interaction is set to reach new heights with the emergence of bio-inspired robots exhibiting more human-like features, movements, and communication abilities. Applications in healthcare, therapy, and social interaction, particularly in contexts like elderly care and special education, will benefit from these developments.(Bensaude-Vincent, n.d.; Neveln et al., n.d.)

Biomimetic design principles geared towards energy efficiency, drawing inspiration from the optimization observed in animals and plants, will become more prevalent. This approach aims to extend the operational capabilities of robots while minimizing energy consumption.(Hayes et al., 2019)

The field of micro-robotics is anticipated to advance, taking cues from microorganisms and enabling new applications, particularly in medicine. Micro-robots, inspired by nature, could perform minimally invasive procedures, revolutionizing medical interventions.(Neveln et al., n.d.)

Environmental sustainability will remain a key focus, with the continued growth of bio-inspired robots dedicated to environmental cleanup, conservation, and monitoring. These robots hold significant potential in addressing urgent global environmental challenges.(Hayes et al., 2019)

The integration of bio-inspired designs into biomechanical prosthetics and exoskeletons is poised to enhance mobility and improve the quality of life for individuals with disabilities. Additionally, these advancements will benefit workers engaged in physically demanding jobs, showcasing the far-reaching impact of biomimetic robotics on human well-being and industry.

Fig.7.4: An AI model depicting robotic crab(*Robot Crab Stock Illustrations – 270 Robot Crab Stock Illustrations, Vectors & Clipart -Dreamstime, n.d.*)

The future of bio-inspired robotics holds exciting prospects with the integration of artificial intelligence (AI) and machine learning (ML). This convergence is poised to elevate bio-inspired robots to new levels of autonomy, decision-making, and adaptability. By incorporating AI and ML, these robots will gain the ability to handle a broader spectrum of tasks within dynamic and complex environments. This shift towards intelligent, learning systems reflects a significant leap forward in the capabilities of bio-inspired robots, enhancing their functionality and versatility.

Moreover, the trajectory of bio-inspired robotics is expected to be shaped by cross-disciplinary collaboration. Partnerships between biologists, engineers, and computer scientists will foster a more holistic approach to the development of robotic systems. This collaborative effort will lead to the creation of robots that not only imitate biological structures but also accurately replicate the intricacies of natural organisms. By bringing together insights from diverse fields, such collaborative endeavors are poised to accelerate progress in bio-inspired robotics, pushing the boundaries of what these robots can achieve.(Bensaude-Vincent, n.d.; Coyle et al., n.d.; Neveln et al., n.d.)

The future of bio-inspired robotics holds great promise for solving complex challenges in various domains. These trends will not only advance technology but also contribute to improving the quality of life, safety, and sustainability in our rapidly evolving world.

Summary

The summary of this chapter based on Bio-Inspired Robotics and Automation highlights the transformative impact of biomimicry across various domains, including search and rescue operations, space exploration, industrial automation, and environmental monitoring. It discusses how biomimetic sensors replicate the acute senses of animals, enhancing search and rescue efforts and planetary exploration. The chapter explores the design and functionality of biomimetic robots, such as gecko-inspired climbers, anthropomorphic robotic hands, and collaborative robots based on social insect behavior. Furthermore, it delves into future trends in bio-inspired robotics, including soft robotics, biohybrid robots, and the integration of artificial intelligence. Overall, the chapter emphasizes the diverse applications and promising future of biomimetic robotics in improving efficiency, safety, and sustainability across industries.

References

Araque, K., Palacios, P., Mora, D., & Austin, M. C. (2021). *Biomimicry-Based Strategies for Urban Heat Island Mitigation: A Numerical Case Study under Tropical Climate.*

Bensaude-Vincent, B. (n.d.). Bio-Informed Emerging Technologies and Their Relation to the Sustainability Aims of Biomimicry. *Environmental*

Values.

Biomimicry, an Approach, for Energy Effecient Building Skin Design. (2016). *Procedia Environmental Sciences.*

Blok, V., & Gremmen, B. (2016). Ecological Innovation: Biomimicry as a New Way of Thinking and Acting Ecologically. *Journal of Agricultural and Environmental Ethics,* 29(2), 203–217. https://doi.org/10.1007/s10806-015-9596-1

Coyle, S., Majidi, C., LeDuc, P., & Hsia, K. J. (n.d.). *Bio-inspired soft robotics: Material selection, actuation, and design.*

Dumanli, A. G., & Savin, T. (2016). Recent advances in the biomimicry of structural colours. *Chemical Society Reviews,* 45(24), 6698–6724. https://pubs.rsc.org/en/content/articlehtml/2008/nd/c6cs00129g

Hayes, S., Desha, C., & Gibbs, M. (2019). *Findings of Case-Study Analysis: System-Level Biomimicry in Built-Environment Design.*

Neveln, I. D., Bai, Y., Snyder, J. B., Solberg, J. R., Curet, O. M., Lynch, K. M., & MacIver, M. A. (n.d.). Biomimetic and bio-inspired robotics in electric fish research. *THE JOURNAL OF EXPERIMENTAL BIOLOGY.*

Perera, A. S., & Coppens, M.-O. (n.d.). *Re-designing materials for biomedical applications: From biomimicry to nature-inspired chemical engineering.*

Robot Crab Stock Illustrations – 270 Robot Crab Stock Illustrations, Vectors & Clipart—Dreamstime. (n.d.). Retrieved April 22, 2024, from https://www.dreamstime.com/illustration/robot-crab.html

Ruiz-Garcia, L. (2009). *A Review of Wireless Sensor Technologies and Applications in Agriculture and Food Industry: State of the Art and Current Trends.*

Steampunk Submarine by LG-Design on DeviantArt. (n.d.). Retrieved April 22, 2024, from https://www.deviantart.com/lg-design/art/Steampunk-Submarine-955519119

Transistor Images | Free Photos, PNG Stickers, Wallpapers & Backgrounds. (n.d.). Rawpixel. Retrieved April 22, 2024, from https://shorturl.at/4PwhW

Turner, A. P. F. (2013). Biosensors: Sense and sensibility. *Chem Soc Rev.*

CHAPTER VIII

Sustainable Energy and Bio Inspiration

This chapter delves into the integration of sustainable energy and bio-inspiration, emphasizing the role of biomimicry in advancing renewable energy technologies. By drawing insights from natural processes and biological systems, researchers aim to optimize energy conversion, storage, and distribution while minimizing environmental impact. Examples of biomimetic energy solutions, ranging from solar and wind power to energy storage systems and architectural design, illustrate the potential of nature-inspired innovation in creating more efficient, resilient, and environmentally friendly energy solutions. The chapter underscores the importance of considering environmental impact and sustainability in energy-related activities, advocating for the adoption of bio-inspired approaches to address the growing global demand for clean and renewable energy sources.

8.1 Renewable Energy and Natural Processes

The intersection of sustainable energy and bio-inspiration involves leveraging natural processes and principles found in the environment to enhance the development and utilization of renewable energy. Here's how renewable energy and bio-inspiration are interconnected:

Bio-inspiration plays a pivotal role in advancing renewable energy technologies, drawing insights from biological systems to optimize energy conversion processes. Examining natural phenomena like photosynthesis in plants provides a blueprint for more efficient energy capture and conversion. By mimicking these intricate processes, bio-inspired renewable energy systems can significantly enhance their performance, making them more adept at harnessing energy from abundant natural sources, such as sunlight. Additionally, biomimetic design principles contribute to the development of adaptive systems capable of responding to fluctuating environmental conditions. This adaptability is particularly beneficial in the realm of variable renewable sources like solar and wind, where the ability to adjust to changing energy availability improves the resilience and reliability of renewable energy systems. Furthermore, the integration of bio-inspired materials stands out as a key strategy. Drawing inspiration from natural structures, such as plant leaves, enables the creation of advanced components for renewable energy technologies. For instance, designing solar panels that replicate the light-absorbing properties of leaves enhances energy harvesting efficiency. In essence, the application of bio-inspired concepts in renewable energy not only drives innovation but also aligns with the efficiency and adaptability observed in nature, paving the way for more sustainable and effective energy solutions.(Chayaamor-Heil & Hannachi-Belkadi, 2017; Imani & Vale, 2022)

Fig.8.1: A solar pannel(*Royalty-Free Photo: Black and Grey Solar Panel | PickPik*, n.d.)

Exploring biomimicry in the realm of renewable energy extends beyond energy conversion to encompass crucial aspects like energy storage, ecosystem integration, resource efficiency, and user-centered design. Nature's blueprint for efficient energy storage, as observed in plants, inspires innovative solutions for storing excess energy generated from renewable sources. Applying biomimicry to energy storage systems holds the potential to develop sustainable and advanced storage solutions that align with the cyclical and efficient energy management found in natural systems.(Amer, 2019a; Araque et al., 2021; Chayaamor-Heil & Hannachi-Belkadi, 2017)

Biomimetic approaches can also revolutionize the integration of energy systems with ecosystems, emphasizing a design philosophy that minimizes environmental impact. By learning from natural processes, designers can create renewable energy solutions that seamlessly coexist with the surrounding environment, promoting sustainability and ecological harmony.(Amer, 2019a, 2019b)

Resource efficiency, a hallmark of natural systems, becomes a guiding principle in biomimetic renewable energy technologies. Drawing inspiration from nature helps design systems that minimize resource consumption throughout their lifecycle—from production and operation to decommissioning. This approach contributes significantly to the overall sustainability of renewable energy solutions.(Chayaamor-Heil & Hannachi-Belkadi, 2017; Imani & Vale, 2022)

Moreover, a user-centered design philosophy, informed by the study of how living organisms interact with their environment, becomes a focal point in bio-inspired renewable energy applications. By understanding human needs and behaviors, bio-inspiration guides the creation of energy solutions that are not only technologically advanced but also align with the user's experience, fostering a more harmonious integration of renewable energy into daily life.

In essence, biomimicry in renewable energy spans a spectrum of considerations, from efficient storage to ecosystem harmony, resource optimization, and user-centric design.

By integrating bio-inspiration into the development of renewable energy technologies, there is potential for more sustainable, efficient , and harmonious solutions that address the growing global demand for clean and renewable energy sources.

8.2 Biomimetic Energy Solutions

Biomimetic energy solutions involve drawing inspiration from biological systems to design and enhance technologies for efficient and environmentally friendly energy production. Here are examples of biomimetic energy solutions:

Researchers are delving into biomimicry to revolutionize renewable energy technologies, with a focus on mimicking natural processes for more efficient energy harvesting. One such innovation involves drawing inspiration from photosynthesis, the process employed by plants to convert sunlight into energy. In this biomimetic approach, scientists are exploring the development of artificial leaves or solar cells that replicate the intricate mechanisms of photosynthesis. The goal is to enhance the efficiency of solar energy capture, contributing to advancements in sustainable energy production.(de Pauw et al., 2010; Quaschning, n.d.)

In the realm of wind energy, biomimicry takes inspiration from the collective behavior of fish in schools. Researchers are exploring how fish schooling behavior can inform the design and arrangement of wind turbines in wind farms. Emulating the coordinated movement of fish in fluid environments holds the promise of optimizing the efficiency of energy capture in wind power, bringing about more effective and environmentally friendly wind energy solutions.(Imani & Vale, 2022; Quaschning, n.d.)

Additionally, biomimetic designs for ocean wave energy harvesting draw inspiration from the flexibility and resilience observed in kelp forests. Engineers are working on structures that mimic the swaying motion of kelp, providing a biomimetic solution for wave energy converters. This approach allows for better adaptation to changing wave patterns, showcasing how insights from nature can lead to more effective and sustainable technologies in the field of renewable energy.(Chayaamor-Heil & Hannachi-Belkadi, 2017; de Pauw et al., 2010).

Fig. 8.2: Biomimicry rainwater harvesting(Withington, 2009)

Researchers are harnessing biomimicry to propel innovations in energy-related domains, seeking inspiration from nature's efficient mechanisms. Biomimicry of plant metabolism, where energy storage and release occur seamlessly, is influencing the design of cutting-edge energy storage systems. This approach aims to replicate the effectiveness observed in plants, paving the way for advanced storage solutions that emulate nature's efficiency in utilizing energy.(Amer, 2019b; Zhang et al., n.d.)

In the realm of architectural sustainability, biomimicry draws inspiration from termite mounds' thermoregulation mechanisms. Biomimetic building cooling systems, inspired by these natural structures, aim to optimize energy efficiency. This involves creating ventilation systems that can maintain comfortable temperatures with minimal energy consumption, showcasing the potential for nature-inspired solutions in sustainable architecture.(Amer, 2019b)

Furthermore, biomimicry extends its reach to the field of energy distribution with beehive-inspired smart grids. Learning from the organized communication and collaboration within beehives, these biomimetic systems emulate the decentralized and adaptive nature of bee colonies. The goal is to enhance the efficiency and resilience of energy distribution networks, introducing a new paradigm inspired by the natural world.(Messner et al., n.d.; Quaschning, n.d.)

In solar energy, researchers are drawing inspiration from the microscopic structures found on butterfly wings. This biomimetic approach is fueling the development of solar concentrators designed to mimic the efficiency of nature. These devices focus sunlight more effectively, offering potential improvements in the performance of concentrated solar power systems. The amalgamation of biomimicry and energy-related innovations showcases how insights from the natural world can lead to sustainable and efficient solutions in diverse technological applications.(El-Zeiny, 2012; Messner et al., n.d.).

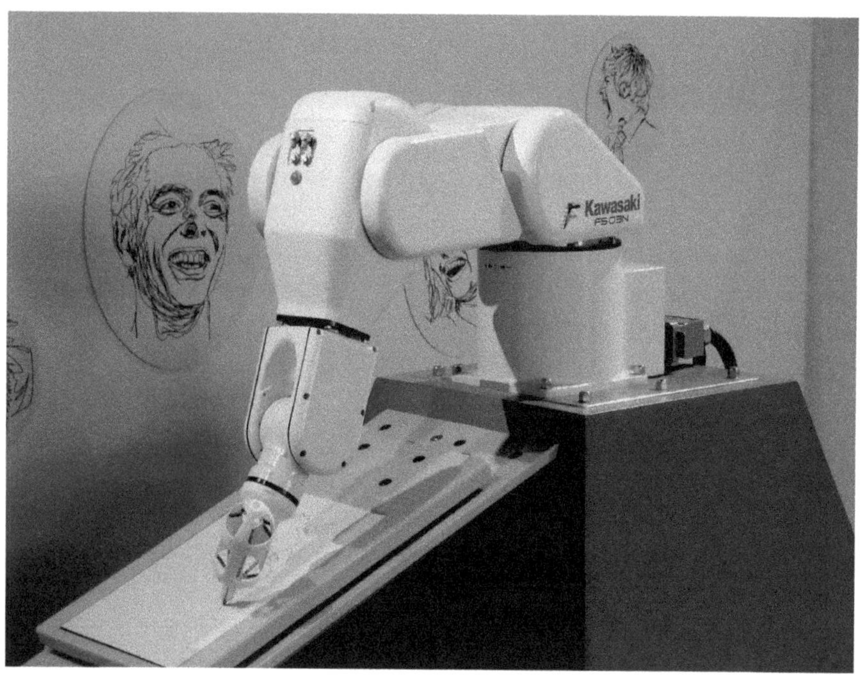

Fig.8.3 A robotic arm(*Download Free Photo of Robot,Robot Arm,Robotics,Painting,Kawasaki - from Needpix.Com*, n.d.)

By integrating principles from the natural world into the design of energy solutions, biomimetic approaches aim to create technologies that not only generate power sustainably but also contribute to the overall resilience and harmony of energy systems with the environment.

8.3 Environmental Impact and Sustainability

The relationship between environmental impact and sustainability is crucial.

Environmental Impact: This refers to the effects that energy-related activities and technologies have on the environment. Traditional energy sources, such as fossil fuels, often result in negative impacts, including air and water pollution, habitat destruction, and greenhouse gas emissions, contributing to climate change.(Bensaude-Vincent, n.d.; Blok & Gremmen, 2016)

Sustainability: In the context of energy, sustainability involves meeting present energy needs without compromising the ability of future generations to meet their own needs. Sustainable energy sources minimize environmental degradation, promote resource efficiency, and contribute to long-term ecological balance.(Chayaamor-Heil & Hannachi-Belkadi, 2017; Quaschning, n.d.)

Bio-Inspiration's Role: Biomimetic approaches in energy solutions aim to reduce environmental impact and enhance sustainability by drawing inspiration from nature's efficient and adaptive processes. This involves creating technologies that integrate seamlessly with ecosystems, optimize resource use, and minimize harm to the environment.(Aziz, n.d.; Benyus, 1997)

By incorporating bio-inspiration into sustainable energy solutions, there is potential to develop technologies that not only reduce environmental impact but also contribute positively to ecological systems, aligning with the principles of sustainability.

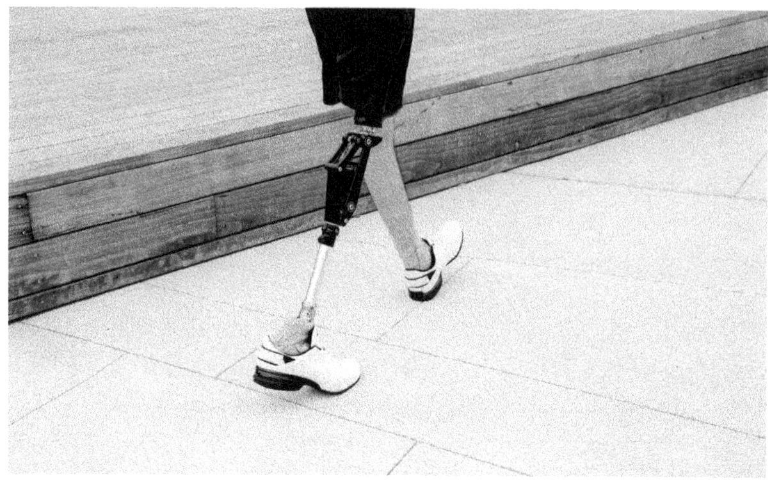

Fig.8.4: A prosthetic leg(*Man Walking with Prosthetic Leg · Free Stock Photo*, n.d.)

Summary

This chapter explores the intersection of sustainable energy and bio-inspiration, highlighting how biomimicry can enhance renewable energy technologies. By emulating natural processes like photosynthesis and collective behavior in biological systems, researchers aim to optimize energy conversion, storage, and distribution. Examples include artificial leaves for solar energy capture, biomimetic wind turbine arrangements, and energy storage systems inspired by plant metabolism. Biomimetic approaches also extend to architectural design, with solutions for energy-efficient buildings inspired by termite mounds and beehive-like smart grids for energy distribution. The chapter emphasizes the importance of reducing environmental impact and promoting sustainability through nature-inspired innovation in energy solutions to meet the global demand for clean and renewable energy sources.

References

Amer, N. (2019a). Biomimetic Approach in Architectural Education: Case study of 'Biomimicry in Architecture' Course. *Ain Shams Engineering Journal.*

Amer, N. (2019b). Biomimetic Approach in Architectural Education: Case study of 'Biomimicry in Architecture' Course. *Ain Shams Engineering Journal.*

Araque, K., Palacios, P., Mora, D., & Austin, M. C. (2021). *Biomimicry-Based Strategies for Urban Heat Island Mitigation: A Numerical Case Study under Tropical Climate.*

Aziz, M. S. (n.d.). *Biomimicry as an approach for bio-inspired structure with the aid of computation.*

Bensaude-Vincent, B. (n.d.). Bio-Informed Emerging Technologies and Their Relation to the Sustainability Aims of Biomimicry. *Environmental Values.*

Benyus, J. M. (1997). *Biomimicry: Innovation inspired by nature.* Morrow New York. https://www.academia.edu/download/5239337/biomimicry-innovation-inspired-by-nature.pdf

Blok, V., & Gremmen, B. (2016). Ecological Innovation: Biomimicry as a New Way of Thinking and Acting Ecologically. *Journal of Agricultural and Environmental Ethics, 29*(2), 203–217. https://doi.org/10.1007/s10806-015-9596-1

Chayaamor-Heil, N., & Hannachi-Belkadi, N. (2017). *Towards a Platform of Investigative Tools for Biomimicry as a New Approach for Energy-Efficient Building Design.*

de Pauw, I., Kandachar, P., Karana, E., Peck, D., & Wever, R. (2010). *NATURE INSPIRED DESIGN: STRATEGIES TOWARDS SUSTAINABILITY.*

Download free photo of Robot,robot arm,robotics,painting,kawasaki—From needpix.com. (n.d.). Retrieved May 9, 2024, from https://shorturl.at/I5nzM

El-Zeiny, R. M. A. (2012). *Biomimicry as a Problem Solving Methodology in Interior Architecture.*

Imani, N., & Vale, B. (2022). *Biomimetic Design for Building Energy Efficiency 2021.*

Man walking with Prosthetic Leg · Free Stock Photo. (n.d.). Retrieved May 9, 2024, from https://www.pexels.com/photo/man-walking-with-prosthetic-leg-9623482/

Messner, W. C., Paik, J., Shepherd, R., Kim, S., & Trimmer, B. A. (n.d.). *Energy for Biomimetic Robots: Challenges and Solutions.*

Quaschning, V. (n.d.). *Understanding Renewable Energy Systems.*

Royalty-Free photo: Black and grey solar panel | PickPik. (n.d.). Retrieved May 9, 2024, from https://shorturl.at/Kp8PU

Withington, D. (2009). *"Elf Shelter" Rainwater Collector* [Photo]. https://www.flickr.com/photos/drewwith/3288999994/

Zhang, M., Gu, Z.-Y., Bosch, M., Perry, Z., & Zhou, H.-C. (n.d.). *Biomimicry in metal–organic materials.*

CHAPTER IX

Biomimicry in Medical and Healthcare Engineering

This chapter explores the profound impact of biomimicry in the field of medical devices and healthcare engineering. It discusses how biomimetic approaches draw inspiration from natural systems to develop innovative solutions, spanning from artificial organs and prosthetics to drug delivery systems and wound healing technologies. By replicating the intricate designs and functions observed in biological organisms, biomimicry contributes to the creation of more effective, patient-friendly, and sustainable medical technologies. Examples of bio-inspired solutions are presented, showcasing the transformative potential of biomimicry in addressing healthcare challenges and advancing medical interventions.

9.1 Medical Devices and Innovations

Biomimicry plays a pivotal role in the realm of medical devices, leveraging insights from biological systems to craft innovative solutions. A prominent application involves the creation of artificial organs modeled after natural structures, where biomimicry is employed to replicate the intricate anatomy and physiological functions of their natural counterparts. Noteworthy examples include artificial hearts and kidneys designed to closely mimic the form and functionality of human organs, showcasing the potential for bio-inspired approaches in addressing organ transplantation challenges.(Perera & Coppens, n.d.)

In the domain of prosthetics, biomimicry principles guide the development of limbs that closely emulate the natural movement of human limbs. This involves the integration of advanced robotics and materials that respond to neural signals, leading to prosthetic limbs capable of providing more natural and intuitive movements for amputees. By drawing inspiration from the sophisticated design of biological limbs, biomimetic prosthetics enhance the overall user experience, contributing to improved functionality and adaptability. These applications underscore how biomimicry in medical device design goes beyond conventional approaches, offering transformative solutions that align with the complexity and efficiency observed in the natural world.

Biomimicry has significantly impacted the field of medical devices, with applications extending to various domains.

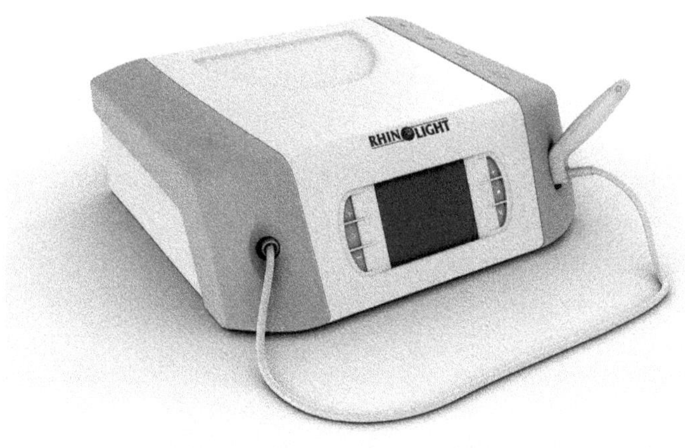

Fig.9.1: An EMF machine(*File*, 2005)

Bio robotics plays a crucial role, where biomimetic robots are engineered to emulate the dexterity and precision found in natural organisms. These robots aid in surgical procedures, enabling minimally invasive surgeries and augmenting surgeons' capabilities. Furthermore, drug delivery systems benefit from biomimicry, as designers draw inspiration from biological systems to create targeted and controlled release mechanisms. This approach enhances drug delivery efficiency while minimizing potential side effects.

In the realm of structural design, biomimetic materials influenced by the structure and properties of bones and tissues are utilized in the development of implants and medical devices. These materials offer a combination of lightweight and durable features, promoting better integration with the human body. Biomimicry is also instrumental in advancing wound-healing technologies, with the creation of bioactive dressings and scaffolds that replicate the natural processes of tissue repair, facilitating accelerated and effective healing.(Bilici et al., n.d.; Perera & Coppens, n.d.)

Inspired by nature's sophisticated filtering systems, biomimetic designs are implemented in medical devices for improved filtration. Taking cues from organs like the kidneys, these designs contribute to the development of enhanced blood filtration devices and artificial organs. Additionally, biomimicry extends to the microscale, where medical devices inspired by the anatomy and behavior of insects are crafted. These microscale technologies find applications in targeted drug delivery and diagnostics, showcasing the efficiency and innovation that biomimicry brings to the forefront of medical advancements.(Bilici et al., n.d.; Budholiya et al., 2021).

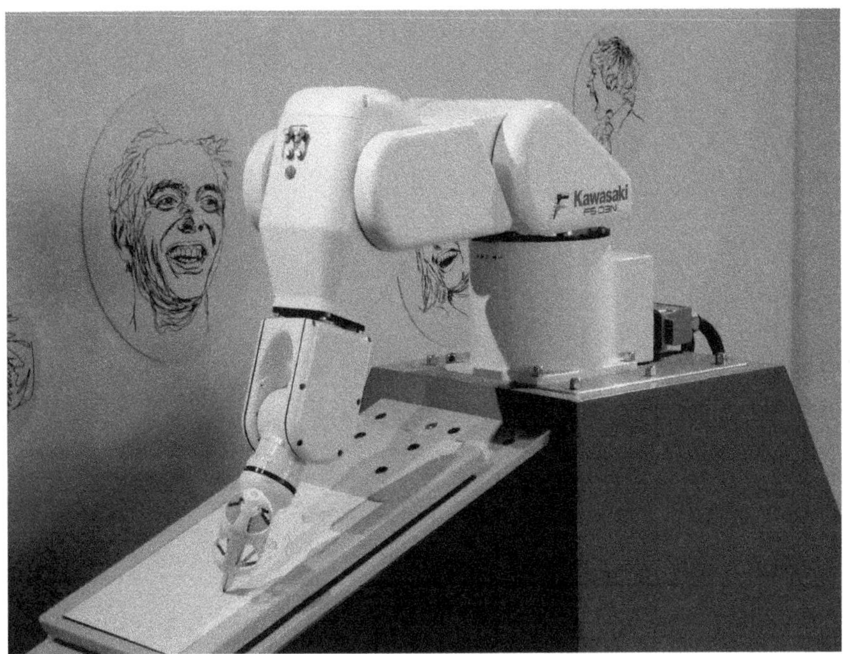

Fig.9.2: A robotic arm(*Download Free Photo of Robot,Robot Arm,Robotics,Painting,Kawasaki - from Needpix.Com, n.d.*)

By applying biomimicry principles to medical and healthcare engineering, researchers aim to develop more effective, patient-friendly, and sustainable solutions that are inspired by the ingenious designs found in the biological world.

9.2 Prosthetics and Biomechanics

In the realm of prosthetics and biomechanics, biomimicry plays a crucial role in designing artificial limbs that closely emulate the natural functions and movements of the human body. Here are key aspects of biomimicry in prosthetics and biomechanics:

Biomimetic prosthetics represent a significant advancement in replicating the natural functionality of human limbs. These innovations go beyond mere replication, aiming to capture the intricate biomechanics and motion observed in natural limbs. Engineers delve into the complexities of muscle, tendon, and joint movements to design prosthetic limbs that emulate a more natural range of motion. Another dimension of biomimicry involves the development of advanced materials inspired by the strength, flexibility, and resilience of muscles and tissues. This approach enhances the comfort and durability of prosthetic limbs, contributing to their overall performance.(Ebeling, n.d.; Majidi, n.d.)

One of the groundbreaking aspects of biomimetic prosthetics is the incorporation of neural interfaces, inspired by the brain's control mechanisms in natural limbs. These interfaces establish a direct connection with neural signals, allowing users to intuitively control the prosthetic limb with their thoughts. This neural interface not only enhances functionality but also significantly improves the user experience, bringing prosthetics closer to mirroring the seamless integration of natural limbs.

Energy efficiency is another key focus of biomimicry in prosthetics. Engineers study how the human body conserves and utilizes energy during movement, aiming to replicate these mechanisms in artificial limbs. The ultimate goal is to create prosthetic limbs that not only require less energy but also offer increased endurance, providing users with a more sustainable and efficient mobility solution. Through these biomimetic approaches, prosthetics are evolving to provide not just physical replacement but a closer approximation to the natural capabilities of human limbs.(Chayaamor-Heil & Hannachi-Belkadi, 2017; Imani & Vale, 2022; Quaschning, n.d.).

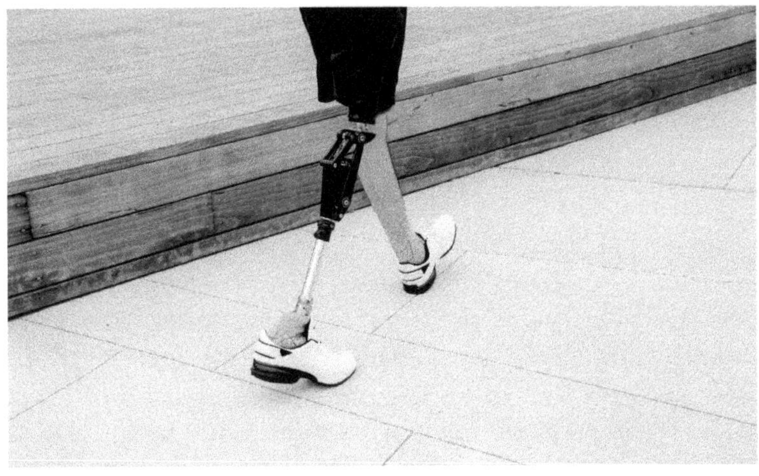

Fig.9.3: A prosthetic leg(*Man Walking with Prosthetic Leg · Free Stock Photo*, n.d.)

Biomimetic prosthetics stand as a profound leap in replicating the intricate functionality of natural human limbs. This transformative innovation transcends mere imitation, aspiring to capture the nuanced biomechanics and fluid motion observed in the complexities of natural limbs. Engineers engage in a meticulous exploration of muscle, tendon, and joint movements to meticulously craft prosthetic limbs that mirror a more authentic range of motion. Another pivotal facet of biomimicry involves the strategic development of advanced materials inspired by the inherent strength, flexibility, and resilience of muscles and tissues. This strategic approach not only augments the comfort and robustness of prosthetic limbs but also significantly elevates their overall performance.(Ebeling, n.d.; Majidi, n.d.)

A groundbreaking dimension of biomimetic prosthetics is the integration of neural interfaces, drawing inspiration from the intricate control mechanisms orchestrated by the human brain in managing natural limbs. These interfaces establish a direct conduit to neural signals, empowering users to intuitively command the prosthetic limb with their thoughts. This neural integration not only amplifies functionality but also profoundly enhances the user experience, bringing prosthetics closer to achieving a seamless emulation of the natural integration of limbs.

Energy efficiency emerges as a paramount focus within the realm of biomimicry in prosthetics. Engineers systematically investigate how the human body conserves and optimizes energy during movement, with the ambition of emulating these physiological mechanisms in artificial limbs. The ultimate objective is to fashion prosthetic limbs that not only demand less energy but also offer heightened endurance, presenting users with a mobility solution that is not only sustainable but also efficient. Through these multifaceted biomimetic strategies, prosthetics are evolving beyond mere physical replacement, endeavoring to provide users with a more authentic approximation of the intricate capabilities inherent in natural human limbs.(Chayaamor-Heil & Hannachi-Belkadi, 2017; Imani & Vale, 2022; Quaschning, n.d.)

Biomimicry in prosthetics and biomechanics aims to not only restore functionality but also enhance the overall user experience by closely emulating the natural capabilities of the human body. This approach contributes to more advanced, comfortable, and user-friendly prosthetic solutions.

9.3 Bio-Inspired Solutions for Healthcare

Biomimicry in medical and healthcare engineering leads to bio-inspired solutions that draw inspiration from nature to address healthcare challenges. Here are examples of bio-inspired solutions:

Biomimicry is reshaping the landscape of medical technologies by drawing inspiration from nature's ingenious solutions. One notable application lies in drug delivery systems that mimic the efficiency of biological transport mechanisms. By designing nanoparticles or carriers that replicate the targeted delivery observed in the human body, this biomimetic approach enhances drug effectiveness while minimizing side effects, representing a significant advancement in medical treatments.(Haque et al., n.d.; Reffad, 2023)

Gecko feet serve as another source of inspiration, influencing the development of medical adhesives with exceptional properties. These bio-inspired adhesives can adhere to wet or moving surfaces, revolutionizing wound closures and surgical adhesives in healthcare settings. The biomimetic design, inspired by nature's adhesive mechanisms, contributes to more effective and versatile medical adhesives.(Bilici et al., n.d.; Iouguina et al., 2014; Reffad, 2023)

The antimicrobial properties of shark skin provide inspiration for the creation of surfaces with innate resistance to bacterial growth. In healthcare settings, these biomimetic surfaces offer a proactive approach to reducing the risk of hospital-acquired infections, addressing a critical concern in medical environments.(Kondoyanni et al., 2022)

Moreover, the strength and flexibility of spider silk have inspired the development of bio-inspired sutures. These sutures, designed to mimic the properties of spider silk, contribute to better wound healing and a decreased risk of complications, showcasing the potential of biomimicry in advancing medical interventions.

In the field of prosthetics, butterfly wing veins provide inspiration for heart valve designs. Biomimicry enhances the durability and flexibility of artificial heart valves, resulting in improved performance and longevity. These bio-inspired designs demonstrate how nature's intricate structures can guide the development of medical solutions with enhanced functionality and e cacy and flexibility of artificial heart valves, improving their performance and longevity.(Iouguina et al., 2014; Majidi, n.d.)

Fig9.4: A butterfly(*Butterfly Wings Veins - Google Search*, n.d.)

Biomimicry in medical technologies continues to flourish, with innovative applications inspired by nature's ingenious solutions. Drawing from the lotus leaf's water-repelling and antimicrobial properties, biomimetic materials are employed to create self-cleaning surfaces in medical equipment. This bio-inspired design not only prevents contamination but also simplifies cleaning procedures, addressing hygiene concerns in healthcare settings.(Perera & Coppens, n.d.)

In the realm of assistive technologies, biomimicry of bat echolocation serves as a source of inspiration for navigation systems tailored to the visually impaired. These bio-inspired solutions utilize sound waves to detect obstacles, providing an enhanced and intuitive means of mobility for individuals with visual impairments.(Iouguina et al., 2014)

Delving into the delicate feeding mechanisms of butterflies, biomimetic hypodermic needles are designed to mimic the long, flexible proboscis of butterflies. This innovative approach seeks to reduce pain and tissue damage during injections, showcasing the potential for bio-inspired designs to improve patient experiences in medical procedures.

The protective scales of silverfish contribute to the development of antibacterial coatings for medical devices. By mimicking the antimicrobial properties observed in silverfish scales, these biomimetic coatings help prevent infections by inhibiting microbial growth on surfaces, adding an extra layer of safety to medical equipment.

Furthermore, the hierarchical vascular network found in plant leaves inspires the design of biomimetic structures for 3D-printed organs. This bio-inspired approach, drawing from leaf vein patterns, enhances the efficiency of nutrient and oxygen transport within artificial tissues, advancing the field of organ printing. These examples underscore how nature's designs serve as a wellspring of inspiration, guiding the development of bio-inspired solutions that address critical challenges in the medical field.(Chayaamor-Heil & Hannachi-Belkadi, 2017, 2017; Imani & Vale, 2022; Iouguina et al., 2014; Majidi, n.d.)

Fig.9.5: Plant leaf veins(*Plant Leaf Veins - Google Search*, n.d.)

Summary

This chapter delves into the application of biomimicry in the realm of medical devices and healthcare engineering. It explores how insights from biological systems are leveraged to craft innovative solutions, such as artificial organs, prosthetics, and drug delivery systems. The chapter highlights the diverse applications of biomimicry, ranging from structural design to assistive technologies, and discusses how these bio-inspired solutions address healthcare challenges while enhancing patient experiences. Through examples and discussions, the chapter underscores the transformative impact of biomimicry in advancing medical interventions and fostering sustainable healthcare practices.

References

Bilici, S. C., Küpeli, M. A., & Guzey, S. S. (n.d.). *Inspired by nature: An engineering design-based biomimicry activity.*

Budholiya, S., Bhat, A., Raj, S. A., Sultan, M. T. H., Shah, A. U., & Basri, A. A. (2021). *State of the Art Review about Bio-Inspired Design and Applications: An Aerospace Perspective.*

butterfly wings veins—Google Search. (n.d.). Retrieved May 9, 2024, from https://shorturl.at/FT4TU

Chayaamor-Heil, N., & Hannachi-Belkadi, N. (2017). *Towards a Platform of Investigative Tools for Biomimicry as a New Approach for Energy-Efficient Building Design.*

Download free photo of Robot,robot arm,robotics,painting,kawasaki—From needpix.com. (n.d.). Retrieved May 9, 2024, from https://www.needpix.com/photo/download/241079/robot-robot-arm-robotics-painting-kawasaki-technology-computer-science-computers

Ebeling, C. (n.d.). *Biomimicry in Mechanical Pros- thesis Design and 3D-printing.*

File:Medical device for ENT-phototherapy.jpg—Wikimedia Commons. (2005, August 18). https://commons.wikimedia.org/wiki/File:Medical_device_for_ENT-phototherapy.jpg

Haque, N. I., Khalil, A. A., Rahman, M. A., Amini, M. H., & Ahamed, S. I. (n.d.). *BIOCAD: Bio-Inspired Optimization for Classification and Anomaly Detection in Digital Healthcare Systems.*

Imani, N., & Vale, B. (2022). *Biomimetic Design for Building Energy Efficiency 2021.*

Iouguina, A., Dawson, J. W., Hallgrimsson, B., & Smart, G. (2014). *BIOLOGICALLY INFORMED DISCIPLINES: A COMPARATIVE ANALYSIS OF BIONICS, BIOMIMETICS, BIOMIMICRY, AND BIO-INSPIRATION AMONG OTHERS. 9*(3).

Kondoyanni, M., Loukatos, D., Maraveas, C., Drosos, C., & Arvanitis, K. G. (2022). *Bio-Inspired Robots and Structures toward Fostering the Modernization of Agriculture.*

Majidi, C. (n.d.). Soft Robotics: A Perspective—Current Trends and Prospects for the Future. *SOFT ROBOTICS*.

Man walking with Prosthetic Leg · Free Stock Photo. (n.d.). Retrieved May 9, 2024, from https://www.pexels.com/photo/man-walking-with-prosthetic-leg-9623482/

Perera, A. S., & Coppens, M.-O. (n.d.). *Re-designing materials for biomedical applications: From biomimicry to nature-inspired chemical engineering.*

plant leaf veins—Google Search. (n.d.). Retrieved May 9, 2024, from https://www.google.com/search?q=plant+leaf+veins&client=ms-android-oppo-

Quaschning, V. (n.d.). *Understanding Renewable Energy Systems.*

Reffad, H. (2023). *A Dynamic Adaptive Bio-Inspired Multi-Agent System for Healthcare Task Deployment. 13*(1).

CHAPTER X

Ethical And Environmental Considerations

This chapter provides a comprehensive overview of ethical and environmental considerations in bio-inspired engineering. It discusses ethical dilemmas, sustainability principles, and responsible bio-inspiration practices. The chapter emphasizes the importance of respecting ecosystems, preserving biodiversity, engaging with communities, and ensuring transparent communication throughout the innovation process. It highlights the integration of ethical frameworks, stakeholder engagement, and sustainable practices to create innovations that positively impact sustainability, conservation, and the well-being of ecosystems and communities.

10.1 Ethical Dilemmas in Bio-Inspired Engineering

Ethical dilemmas in bio-inspired engineering arise from the intersection of innovation and the potential impact on living organisms and ecosystems. Some key ethical considerations include:

The ethical implications of biomimicry encompass several critical dimensions that demand careful consideration. Firstly, there is a concern about intellectual property and fair compensation for the organisms or ecosystems that inspire bio-inspired designs. The replication of biological structures or processes raises ethical questions when it occurs without proper acknowledgment or remuneration to the living entities providing the inspiration. This ethical dilemma necessitates a thoughtful exploration of fair and responsible use of biological knowledge, ensuring that innovation is coupled with ethical considerations and respect for the source of inspiration.(Blok & Gremmen, 2016; Ilieva et al., 2022a)

Secondly, the impact on ecosystems is a vital ethical aspect to contemplate. Extracting materials or disrupting ecosystems for bio-inspired designs can have unintended and potentially harmful consequences on biodiversity and ecological balance. Ethical considerations involve a delicate balance, requiring a thorough assessment of the benefits of innovation against the potential harm to natural habitats. This ethical framework seeks to minimize adverse impacts on ecosystems while advancing technological innovations.(Benyus, 1997; Blok, n.d.)

Moreover, ethical bio-inspired engineering emphasizes responsible biomimicry. It underscores the importance of developing technologies that do not harm or exploit living organisms, fostering a harmonious relationship between innovation and ethical treatment of the environment. This approach acknowledges the interconnectedness of technological progress and environmental responsibility, emphasizing the need for ethical considerations in the development and application of bio-inspired technologies.

Lastly, ethical dilemmas arise in the realm of animal testing for biomimetic devices, particularly in the context of bio-inspired medical advancements. The development and testing of such devices may involve animal testing, prompting ethical questions about the welfare and rights of animals used in research. This emphasizes the imperative to explore alternatives to animal testing or, when unavoidable, to adhere to ethical and humane practices, minimizing harm and prioritizing the well-being of the animals involved. Ethical considerations, therefore, play a pivotal role in guiding the responsible and conscientious development of biomimetic technologies.(Blok, n.d.; Blok & Gremmen, 2016)

Ethical considerations in biomimicry extend into diverse areas, including human research, ecological impact, access to innovations, and the consequences of biomimetic materials. In the realm of human biomimicry research, the principle of informed consent takes center stage. Ethical practices dictate that, particularly in healthcare-related bio-inspired solutions, obtaining informed consent from human participants in research studies is imperative. This ensures that individual autonomy is respected, and transparency is maintained throughout the research process, adhering to fundamental ethical principles.(Blok & Gremmen, 2016)

Fig.10.1: Spider mouth inspired robotic arm(*From the Deep Sea to Deep Space*, n.d.)

Another ethical dimension revolves around the potential unintended consequences of bioengineered organisms introduced into ecosystems. This scenario poses ethical concerns about potential ecological disruptions, the spread of modified genes, and the long-term impact on biodiversity. As bioengineered organisms are employed for various purposes, ethical considerations underscore the importance of comprehensive risk assessments to mitigate unintended ecological consequences.

Equity emerges as a central ethical consideration in the realm of bio-inspired innovations. The ethical lens is focused on ensuring equitable access to the benefits derived from bio-inspired engineering. Questions surrounding affordability, accessibility, and distribution, especially in the context of healthcare solutions originating from biomimicry, prompt a careful examination of how these innovations are made available to diverse populations.

Finally, the use of biomimetic materials across applications raises ethical questions about their potential long-term effects on both human health and the environment. Ethical bio-inspired engineering necessitates thorough testing and consideration of potential consequences to safeguard against adverse impacts. In this way, ethical considerations permeate various facets of biomimicry, guiding the responsible development and application of bio-inspired technologies.(Amer, 2019; Deldin & Schuknecht, n.d.)

Ethical considerations in bio-inspired engineering extend to the realm of cultural and indigenous knowledge, introducing complex dilemmas. When bio-inspired engineering incorporates traditional knowledge or indigenous practices, ethical challenges arise concerning the respectful and equitable use of such cultural contributions. Striking a balance between leveraging valuable insights and preventing exploitation becomes a crucial ethical consideration. Respecting the originators of the knowledge, acknowledging their contributions, and avoiding any form of appropriation are central tenets in navigating these ethical dimensions.(Blok & Gremmen, 2016; Ilieva et al., 2022b; Verbrugghe et al., 2023)

Transparency and effective communication represent fundamental pillars of ethical bio-inspired engineering. Ensuring openness in all stages, from research and development to the communication of potential risks and benefits, is imperative. Ethical practices necessitate engaging in open dialogues with various stakeholders, including the public. This transparency fosters an environment of trust and accountability, allowing for ethical decision-making throughout the innovation process. By prioritizing transparency, ethical bio-inspired engineering strives to uphold the principles of fairness, responsibility, and respect in its engagement with cultural knowledge and its communication with the broader public.

Addressing these ethical dilemmas involves incorporating ethical frameworks, stakeholder engagement, and a commitment to responsible and sustainable bio-inspired engineering practices.

10.2 Sustainability and Conservation

When considering sustainability and conservation in the context of innovation and technology, several ethical and environmental considerations come to the forefront:

Ethical bio-inspired engineering places a strong emphasis on addressing environmental concerns associated with resource consumption, depletion, and the overall lifecycle of technologies. One pivotal aspect involves minimizing the depletion of natural resources, aligning with sustainable practices. Ethical considerations extend to the responsible extraction of materials for bio-inspired engineering, emphasizing responsible resource management to curb adverse environmental effects.(Blok & Gremmen, 2016; Ilieva et al., 2022b; Verbrugghe et al., 2023)

A critical facet of ethical bio-inspired engineering is the comprehensive evaluation of the entire lifecycle of technologies. Conducting a life cycle assessment becomes imperative, ensuring ethical considerations are incorporated from production to disposal. This approach involves a meticulous examination of the environmental impact throughout the technology's existence, guiding the promotion of designs that actively minimize negative effects on ecosystems and natural processes.(Aziz, n.d.; Benyus, 1997; Bilici et al., n.d.-a)

Furthermore, the conservation of biodiversity emerges as a central ethical principle. Ethical bio-inspired engineering actively prioritizes the preservation of biodiversity, acknowledging the interconnectedness of ecosystems. This involves a thorough assessment of potential impacts on local flora and fauna when implementing bio-inspired solutions. By avoiding practices that could contribute to species decline or habitat disruption, ethical bio-inspired engineering seeks to contribute positively to the delicate balance of the natural world. These ethical considerations underscore the commitment to environmental stewardship and responsible innovation in the realm of bio-inspired engineering.(Amer, 2019; Bilici et al., n.d.-a)

Ethical bio-inspired engineering takes a multifaceted approach to environmental and social considerations. Ensuring the resilience of ecosystems is a fundamental principle, with ethical bio-inspired technologies designed to enhance ecological health and diversity without compromising the stability of ecosystems. A central ethical tenet involves active participation in climate change mitigation efforts. Innovations within bio-inspired engineering should strive to reduce greenhouse gas emissions, improve energy efficiency, and facilitate the transition to sustainable energy sources, contributing to global environmental sustainability.

Environmental justice is paramount in ethical bio-inspired engineering, requiring careful consideration to prevent the disproportionate impact of technologies on marginalized communities. Waste reduction and recycling strategies form an integral part of ethical practices, advocating for the use of recyclable materials and minimizing the environmental footprint of waste disposal. In the agricultural context, ethical bio-inspired engineering promotes sustainable farming practices, emphasizing the importance of soil health, biodiversity preservation, and the reduction of harmful chemicals in agricultural ecosystems.(Bensaude-Vincent, n.d.; Blok & Gremmen, 2016)

Beyond environmental considerations, ethical bio-inspired engineering encompasses an evaluation of the social and cultural impacts of innovations. This involves engaging with local communities, respecting cultural diversity, and ensuring that technologies align with, rather than disrupt, local traditions and values. Equitable access to sustainable technologies is another ethical imperative, addressing issues of affordability, accessibility, and inclusivity to ensure global benefit-sharing.(Bensaude-Vincent, n.d.; Blok & Gremmen, 2016; Hayes et al., 2019)

Community engagement and informed consent are critical ethical considerations, particularly when bio-inspired projects have implications for local environments and communities. This commitment to ethical principles underscores the broader goal of ethical bio-inspired engineering — not only to innovate responsibly but to actively contribute to a sustainable and harmonious relationship between technology and the environment, with positive social impacts.

Fig.10.2: A village pond(*Filling Water from Pond in Village - Google Search*, n.d.)

By integrating these ethical and environmental considerations into the development and implementation of bio-inspired technologies, it becomes possible to create innovations that not only address challenges but also contribute positively to sustainability, conservation, and the well-being of ecosystems and communities.

10.3 Responsible Bio-Inspiration

"Responsible bio-inspiration" involves ethical and environmentally conscious practices in deriving inspiration from nature for engineering and innovation. Here are key considerations: Responsible bio-inspiration stands on the foundational principle of respecting and preserving ecosystems, emphasizing a commitment to ecological well-being. This commitment involves conducting comprehensive environmental impact assessments, ensuring a thorough understanding of the potential consequences of bio-inspired projects. The aim is to contribute positively to the health of ecosystems rather than inadvertently causing harm. Ethical considerations play a pivotal role in the responsible sourcing of biological inspiration. This entails obtaining proper permissions and conducting studies with the utmost respect for the organisms involved. Practices that could harm or exploit living organisms are actively avoided, reflecting a dedication to ethical standards in bio-inspired engineering(Benyus, 1997; Bilici et al., n.d.-a; Hayes et al., 2019).

Furthermore, responsible bio-inspiration aligns with the goal of contributing to the conservation of biodiversity. This aspect involves a thoughtful examination of the potential impact that bio-inspired projects may have on local flora and fauna. The emphasis is on minimizing negative consequences and actively participating in the preservation of species diversity. Sustainable material selection is a critical component of responsible bio-inspired engineering. This involves the careful choice of materials with minimal environmental impact, promoting resource efficiency, and considering the entire lifecycle of the materials utilized. In essence, responsible bio-inspiration seeks to harmonize innovation with ecological preservation, fostering a balance that supports both technological advancement and environmental sustainability.(Aziz, n.d.; Benyus, 1997; Bilici et al., n.d.-b).

Fig.10.3: Buterfly sitting on a flower
(*Flora and Fauna - Google Search*, n.d.)

Responsible bio-inspiration is deeply rooted in cultural sensitivity, recognizing and respecting the rich tapestry of cultural diversity. It involves a commitment to acknowledging and valuing indigenous knowledge, traditions, and practices, while actively avoiding any form of appropriation or exploitation of cultural resources. This ensures that bio-inspired projects are developed with a nuanced understanding of cultural contexts and a commitment to fostering positive collaborations.

Transparent communication is a cornerstone of ethical bio-inspiration, encompassing clear articulation of project goals, methods, and potential impacts. This transparency extends to engaging stakeholders, including the public, in a meaningful dialogue that ensures informed decision-making. Environmental justice is a key concern, with responsible bio-inspiration actively addressing potential differential impacts on various communities to avoid disproportionately burdening marginalized or vulnerable populations.(Bilici et al., n.d.-b)

Furthermore, regulatory compliance is a fundamental aspect of responsible bio-inspiration. Adhering to applicable laws and regulations involves obtaining necessary permits, following ethical guidelines, and ensuring compliance with environmental protection standards. Evaluating the social and economic impact of bio-inspired projects is another crucial dimension, considering how innovations may affect communities, employment, and overall well-being. The goal is to ensure positive contributions to society and equitable distribution of benefits.

Education and outreach play a pivotal role in ethical bio-inspiration, with responsible practices involving efforts to educate the public and stakeholders about bio-inspired projects. This includes outreach initiatives designed to promote understanding, address concerns, and actively involve the community in decision-making processes. In essence, responsible bio-inspiration strives for a holistic and ethical approach that integrates cultural awareness, transparent communication, environmental justice, regulatory compliance, and positive social and economic impact.(Aziz, n.d.; Benyus, 1997; Bilici et al., n.d.-b)

By embracing responsible bio-inspiration, innovators and engineers can ensure that their projects not only advance technological solutions but also contribute to the well-being of the environment, respect cultural diversity, and prioritize ethical practices throughout the design and implementation process.

Fig10.4: Dog inpired biomemetic robot
(*Bio Inspired Robot - Google Search*, n.d.)

Summary

This chapter encapsulates the key points discussed within the chapter, providing a concise overview of its contents. It highlights the ethical dilemmas inherent in bio-inspired engineering, including concerns regarding intellectual property, ecosystem impact, and animal testing. Additionally, the summary outlines the ethical principles of responsible biomimicry, emphasizing the importance of ethical sourcing, cultural sensitivity, and transparency in communication. It also discusses the environmental considerations of sustainability and conservation, focusing on responsible resource management, biodiversity preservation, and environmental justice. Lastly, the summary touches upon the concept of responsible bio-inspiration, which integrates ethical, environmental, and cultural considerations into the development and implementation of bio-inspired technologies.

References

Amer, N. (2019). Biomimetic Approach in Architectural Education: Case study of 'Biomimicry in Architecture' Course. *Ain Shams Engineering Journal.*

Aziz, M. S. (n.d.). *Biomimicry as an approach for bio-inspired structure with the aid of computation.*

Bensaude-Vincent, B. (n.d.). Bio-Informed Emerging Technologies and Their Relation to the Sustainability Aims of Biomimicry. *Environmental Values.*

Benyus, J. M. (1997). *Biomimicry: Innovation inspired by nature.* Morrow New York. https://www.academia.edu/download/5239337/biomimicry-innovation-inspired-by-nature.pdf

Bilici, S. C., Küpeli, M. A., & Guzey, S. S. (n.d.-a). *Inspired by nature: An engineering design-based biomimicry activity.*

Bilici, S. C., Küpeli, M. A., & Guzey, S. S. (n.d.-b). *Inspired by nature: An engineering design-based biomimicry activity.*

bio inspired robot—Google Search. (n.d.). Retrieved May 9, 2024, from https://picryl.com/media/bio-inspired-big-dog-quadruped-robot-is-being-developed-as-a-mule-that-can-3a8168

Blok, V. (n.d.). *Ecological Innovation: Biomimicry as a New Way of Thinking and Acting Ecologically.*

Blok, V., & Gremmen, B. (2016). Ecological Innovation: Biomimicry as a New Way of Thinking and Acting Ecologically. *Journal of Agricultural and Environmental Ethics,* 29(2), 203–217. https://doi.org/10.1007/s10806-015-9596-1

Deldin, J.-M., & Schuknecht, M. (n.d.). *The AskNature Database: Enabling Solutions in Biomimetic Design.*

filling water from pond in village—Google Search. (n.d.). Retrieved May 9, 2024, from https://shorturl.at/wuH3a

flora and fauna—Google Search. (n.d.). Retrieved May 9, 2024, from https://www.google.com/search?q=flora+and+fauna+&client=ms-android-oppo-

From the Deep Sea to Deep Space: Sea Urchin's Teeth Inspire Design … (n.d.). Www.Google.Com. Retrieved May 9, 2024, from https://www.designworldonline.com/from-the-deep-sea-to-deep-space-sea-urchins-teeth-inspire-design-for-space-exploration-device/

Hayes, S., Desha, C., & Gibbs, M. (2019). *Findings of Case-Study Analysis: System-Level Biomimicry in Built-Environment Design.*

Ilieva, L., Ursano, I., Traista, L., Hoffmann, B., & Dahy, H. (2022a). *Biomimicry as a Sustainable Design Methodology—Introducing the 'Biomimicry for Sustainability' Framework.*

Ilieva, L., Ursano, I., Traista, L., Hoffmann, B., & Dahy, H. (2022b). *Biomimicry as a Sustainable Design Methodology—Introducing the 'Biomimicry for Sustainability' Framework.*

Verbrugghe, N., Rubinacci, E., & Khan, A. Z. (2023). *Biomimicry in Architecture: A Review of Definitions, Case Studies, and Design Methods.*

CHAPTER XI

Future Horizons in Bio-Inspired Engineering

The chapter explores the future horizons, research opportunities, challenges, and potential breakthroughs in bio-inspired engineering. It highlights the interdisciplinary nature of the field, encompassing medicine, robotics, materials science, energy, and environmental applications. The abstract discusses the emergence of biohybrid systems, advancements in neural interfaces, environmental monitoring technologies, energy solutions, and transformative innovations in healthcare. It also addresses challenges such as ethical considerations, interdisciplinary integration, regulatory frameworks, and resource limitations. Despite challenges, the chapter underscores the potential for groundbreaking breakthroughs in various domains through responsible innovation and collaboration in bio-inspired engineering.

11.1 Emerging Technologies and Fields

"Future Horizons in Bio-Inspired Engineering" signify the dynamic and expansive frontiers within the realm of applying biological principles to engineering, envisioning emerging technologies and new fields of study. This forward-looking perspective encompasses a spectrum of key aspects.(Amer, 2019)

Firstly, there is a focus on innovative applications spanning diverse industries such as medicine, robotics, materials science, and energy. These applications draw inspiration from biological systems, promising groundbreaking advancements.

Secondly, the trajectory of bio-inspired engineering involves extensive interdisciplinary collaborations, where researchers from biology, engineering, chemistry, and other fields collaborate to forge holistic and synergistic solutions.

The third aspect revolves around the evolution of robotics, with an emphasis on developing sophisticated and adaptable robots inspired by animal locomotion, sensing capabilities, and swarm intelligence.

Fig.11.1: Nature and biodiversity(*Captivating Nature: Biodiversity, Conservation & Eco-Friendly Tips | AI Art Generator | Easy-Peasy.AI*, n.d.)

Another significant dimension is the integration of synthetic biology with bio-inspired engineering, paving the way for biohybrid systems that combine living and synthetic components for enhanced functionalities. Additionally, there is a strong emphasis on neuroengineering, aiming to mimic and comprehend the intricacies of the human brain, potentially leading to breakthroughs in brain-machine interfaces, neuroprosthetics, and cognitive computing. Anticipated applications extend to environmental monitoring and remediation, with bio-inspired technologies offering innovative solutions for pollution detection, air and water purification, and sustainable environmental management.(Amer, 2019; *Captivating Nature: Biodiversity, Conservation & Eco-Friendly Tips | AI Art Generator | Easy-Peasy.AI*, n.d.; Hayes et al., 2019)

The emergence of biologically augmented materials, including lightweight and strong materials, self-healing polymers, and adaptive surfaces, holds promise for applications in aerospace, construction, and consumer goods.

Furthermore, bio-inspired solutions for energy harvesting and storage are anticipated to play a significant role, leveraging insights from nature's e ciency in energy conversion. In the healthcare domain, bio-inspired engineering is expected to drive transformative innovations in medical devices, drug delivery systems, tissue engineering, and personalized medicine.(Hwang et al., n.d.; Kong et al., 2023).

Fig.11.2: A tribal community(*USAID Measuring Impact Conservation Enterprise Retrospecti... | Flickr*, n.d.)

Lastly, the future of bio-inspired engineering emphasizes education and ethical considerations, underlining the importance of educating professionals and the public about this field. Ethical considerations will be paramount to ensure responsible and sustainable practices in the development and deployment of bio-inspired technologies. Collectively, these future horizons underscore the exciting possibilities and potential breakthroughs in bio-inspired engineering, highlighting how nature's ingenious solutions continue to inspire transformative technologies to address complex challenges across diverse domains. (Humenik et al., 2011; Hwang et al., n.d.)

11.2 Research and Innovation Opportunities

The future horizons in bio-inspired engineering open up extensive avenues for research and Normalinnovation across diverse domains. Researchers and innovators can delve into several key areas to explore new frontiers. First and foremost, there is a compelling opportunity to investigate the development of advanced biomaterials inspired by natural structures and properties. (Ersanlı & Ersanlı, n.d.; Humenik et al., 2011).

This encompasses the exploration of materials with heightened strength, flexibility, and adaptability, potentially revolutionizing applications in medicine, construction, and consumer goods. Another promising avenue is in the realm of neural interfaces and cognitive technologies.

Advancements in neuroengineering can be pursued by developing sophisticated neural interfaces and cognitive technologies, with a focus on researching brain-machine interfaces, neuroprosthetics, and cognitive computing to enhance human-machine interactions and aid individuals with neurological conditions. The field of robotics offers a platform for innovation by integrating biological principles to augment robot capabilities. (*Human Brain Neurons: Understanding the Intricacies of Neural Networks | AI Art Generator | Easy-Peasy.AI*, n.d.).

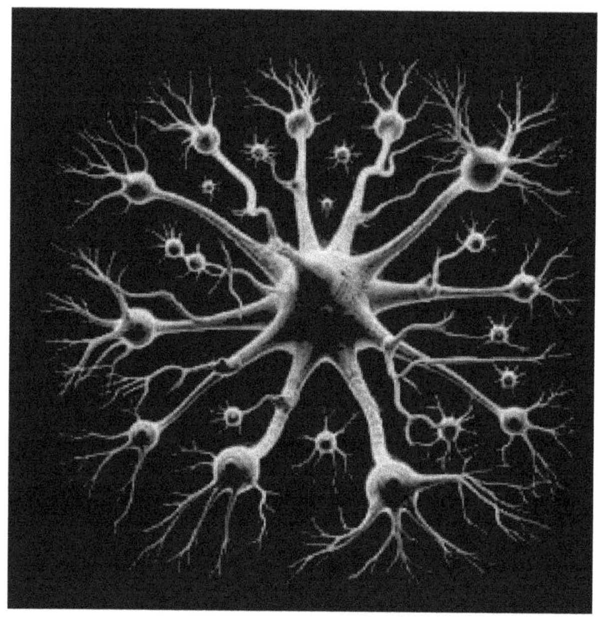

Fig.11.3: An AI image depicting human brain nauron(*Human Brain Neurons: Understanding the Intricacies of Neural Networks | AI Art Generator | Easy-Peasy.AI, n.d.*)

This involves designing robots inspired by animal behaviors, sensory systems, and swarm intelligence, with potential applications in healthcare, exploration, and manufacturing. Additionally, there is an opportunity to explore the integration of synthetic biology with bio-inspired engineering, leading to the creation of biohybrid systems that combine living and synthetic components, fostering innovations like bio-hybrid robots or biocompatible electronics.

Fig11.4: An AI image of tree surrounded by wooden chairs and table(*Tree-Encircled Seating | Serene Outdoor Setting | AI Art Generator | Easy-Peasy.AI, n.d.*)

Environmental monitoring and remediation technologies represent a crucial area where bio-inspired solutions can be developed. Researchers can investigate technologies inspired by natural systems, such as plants or microorganisms, to address pollution detection, air and water purification, and sustainable environmental management. Energy harvesting and storage solutions offer another exciting prospect for exploration, involving bio-inspired approaches that mimic nature's e ciency in energy conversion, with the aim of developing advanced and sustainable energy solutions.(Aslan et al., 2022; Boghossian et al., 2011; Hayes et al., 2019).

The healthcare sector presents opportunities for transformative innovations inspired by biological systems. Researchers can focus on advancements in medical devices, drug delivery systems, tissue engineering, and personalized medicine to enhance patient outcomes and healthcare e ciency. Educational initiatives play a vital role in increasing awareness and understanding of bio-inspired engineering. (Haque et al., n.d.; Reffad, 2023).

This includes the development of educational programs that foster interdisciplinary learning and promote the ethical and responsible application of bio-inspired technologies. Encouraging cross-disciplinary collaborations between researchers in biology, engineering, chemistry, and related fields is essential for holistic advancements. Establishing ethical frameworks and guidelines for bio-inspired engineering research and innovation is imperative, addressing considerations related to responsible sourcing, environmental impact, and societal implications (Amer, 2019).

Lastly, contributing to global sustainability initiatives is a collective responsibility, and aligning bio-inspired engineering research with environmental and societal sustainability goals can play a significant role. In summary, the future horizons in bio-inspired engineering offer a diverse and rich landscape for researchers and innovators to explore transformative advancements. Spanning biomaterials, robotics, neuroscience, environmental technologies, energy solutions, healthcare, education, and ethical considerations, these opportunities present a chance to draw inspiration from nature and address complex challenges while contributing to the well-being of ecosystems and societies (Boghossian et al., 2011; Hayes et al., 2019).

11.3 Challenges and Potential Breakthroughs

The future of bio-inspired engineering unfolds a landscape marked by both formidable challenges and the promise of revolutionary breakthroughs. Effectively navigating these challenges while harnessing the vast opportunities for innovation is imperative. The intricate complexity inherent in biological systems presents a formidable challenge, demanding a profound understanding of biological intricacies and the development of sophisticated modeling and simulation techniques to accurately replicate and integrate these systems into engineered solutions. Interdisciplinary integration, a fundamental aspect of bio-inspired engineering, necessitates effective collaboration and communication among diverse disciplines such as biology, engineering, and materials science, to foster a holistic approach where insights from various fields synergize for innovative solutions (Ersanlı & Ersanlı, n.d.).

As the field progresses, ethical considerations become paramount, encompassing responsibly sourcing biological inspiration, addressing environmental and societal impacts, and establishing guidelines to ensure the ethical use of bio-inspired technologies. Bridging the gap between theoretical concepts and practical applications remains a persistent challenge, requiring the overcoming of technical, logistical, and scalability obstacles to seamlessly integrate bio-inspired ideas into tangible technologies applicable in real-world scenarios. Ensuring the adaptability and robustness of bio-inspired technologies in dynamic environments proves to be another challenge, demanding innovative design approaches and materials that mimic the resilience and adaptability observed in natural systems (Favi et al., 2014).

Resource limitations, both in terms of funding and raw materials, can impede progress, necessitating efforts to secure sustainable funding and explore alternative materials and technologies. The introduction of bio-inspired technologies may lead to unintended consequences, emphasizing the need to anticipate and mitigate potential negative impacts such as ecological disruptions or unforeseen side effects for responsible innovation. Public perception and acceptance present challenges that must be addressed by effectively communicating the benefits and fostering public understanding to garner support and trust in these emerging technologies (Aïssa, 2014; Broeckhoven & Du Plessis, 2022).

The absence of comprehensive regulatory frameworks for bio-inspired engineering stands as a significant impediment, highlighting the need to establish clear guidelines and regulations ensuring the safety, ethical use, and environmental sustainability of bio-inspired technologies. Ensuring the long-term sustainability of bio-inspired solutions is a multifaceted challenge, involving considerations of environmental impacts, lifecycle of materials, and the development of technologies that withstand the test of time. (Broeckhoven & Du Plessis, 2022).

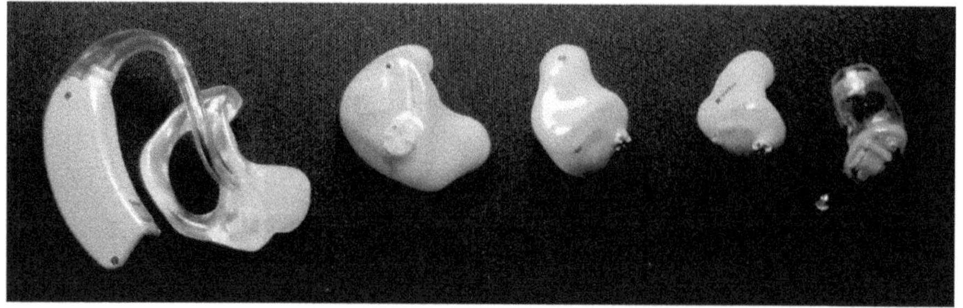

Fig.11.5: Ear sensor for enhansing listening(ikesters, 2012)

Despite these challenges, the future holds the potential for groundbreaking breakthroughs. Bio-inspired innovations may lead to transformative advancements in medical fields, including personalized medicine, organ regeneration, and advanced prosthetics. Contributions to environmental stewardship are expected, with bio-inspired solutions providing sustainable alternatives for energy production, waste management, and pollution control. Breakthroughs in materials science inspired by nature could result in the development of smart and adaptive materials with applications in construction, aerospace, and consumer goods.

Summary

The chapter outlines the future prospects, research avenues, challenges, and potential breakthroughs in bio-inspired engineering. It emphasizes interdisciplinary collaboration across medicine, robotics, materials science, energy, and environmental fields. Key areas of exploration include biomaterials, neural interfaces, robotics, environmental technologies, energy solutions, and healthcare innovations. Challenges such as ethical considerations, interdisciplinary integration, regulatory frameworks, and resource limitations are discussed. Despite these challenges, the chapter underscores the potential for transformative advancements through responsible innovation and collaboration in the dynamic field of bio-inspired engineering.

Advancements in robotics, incorporating biological principles, may lead to the development of robots with unprecedented adaptability, agility, and intelligence. Bio-inspired neuroengineering holds promise for breakthroughs in understanding and treating neurological conditions, as well as enhancing human-machine interfaces. Additionally, breakthroughs in educational initiatives could lead to increased public awareness, understanding, and acceptance of bio-inspired engineering, fostering a more informed and supportive global community. Navigating these challenges and capitalizing on breakthrough opportunities requires sustained collaboration, ethical considerations, and a steadfast commitment to responsible innovation in the dynamic and evolving field of bio-inspired engineering.

References

Aïssa, B. (2014). *Self-healing Materials: Innovative Materials for Terrestrial and Space Applications*. Smithers Rapra.

Amer, N. (2019). Biomimetic Approach in Architectural Education: Case study of 'Biomimicry in Architecture' Course. *Ain Shams Engineering Journal*.

Aslan, D., Selçuk, S. A., & Avinç, G. M. (2022). A Biomimetic Approach to Water Harvesting Strategies: An Architectural Point of View. *International Journal of Built Environment and Sustainability*, 9(3), 47–60. https://ijbes.utm.my/index.php/ijbes/article/view/969

Boghossian, A. A., Ham, M.-H., Choi, J. H., & Strano, M. S. (2011). Biomimetic strategies for solar energy conversion: A technical perspective. *Energy & Environmental Science*, 4(10), 3834–3843. https://pubs.rsc.org/en/content/articlehtml/2011/ee/c1ee01363g

Broeckhoven, C., & Du Plessis, A. (2022). Escaping the Labyrinth of Bioinspiration: Biodiversity as Key to Successful Product Innovation. *Advanced Functional Materials*, 32(18), 2110235. https://doi.org/10.1002/adfm.202110235

Captivating Nature: Biodiversity, Conservation & Eco-friendly Tips | AI Art Generator | Easy-Peasy.AI. (n.d.). Retrieved May 9, 2024, from https://easy-peasy.ai/ai-image-generator/images/diverse-wildlife-habitat-seasonal-changes-conservation-awareness

Ersanlı, E. T., & Ersanlı, C. C. (n.d.). Biomimicry: Journey to the Future with the Power of Nature. *International Scientific and Vocational Studies Journal*, 7(2), 149–160. Retrieved May 15, 2024, from https://dergipark.org.tr/en/pub/bilmes/issue/82358/1388402

Favi, P. M., Yi, S., Lenaghan, S. C., Xia, L., & Zhang, M. (2014). Inspiration from the natural world: From bio-adhesives to bio-inspired adhesives. *Journal of Adhesion Science and Technology*, 28(3–4), 290–319. https://doi.org/10.1080/01694243.2012.691809

Haque, N. I., Khalil, A. A., Rahman, M. A., Amini, M. H., & Ahamed, S. I. (n.d.). *BIOCAD: Bio-Inspired Optimization for Classification and Anomaly Detection in Digital Healthcare Systems.*

Hayes, S., Desha, C., & Gibbs, M. (2019). *Findings of Case-Study Analysis: System-Level Biomimicry in Built-Environment Design.*

Human Brain Neurons: Understanding the Intricacies of Neural Networks | AI Art Generator | Easy-Peasy.AI. (n.d.). Retrieved May 9, 2024, from https://shorturl.at/As432

Humenik, M., Smith, A. M., & Scheibel, T. (2011). Recombinant spider silks—Biopolymers with potential for future applications. *Polymers*, 3(1), 640–661. https://www.mdpi.com/2073-4360/3/1/640

Hwang, J., Jeong, Y., Park, J. M., Lee, K. H., Wook, J., & Choi, J. (n.d.). Biomimetics: Forecasting the future of science, engineering, and medicine. *International Journal of Nanomedicine.*

ikesters. (2012). *English: Traditional hearing aid designs.* https://commons.wikimedia.org/wiki/ File:Traditional_hearing_aids.jpg

Kong, B., Qi, C., Wang, H., Kong, T., & Liu, Z. (2023). Tissue adhesives for wound closure. *Smart Medicine*, 2(1), e20220033. https://doi.org/ 10.1002/SMMD.20220033

Reffad, H. (2023). *A Dynamic Adaptive Bio-Inspired Multi-Agent System for Healthcare Task Deployment. 13*(1).

Tree-Encircled Seating | Serene Outdoor Setting | AI Art Generator | Easy-Peasy.AI. (n.d.). Retrieved May 9, 2024, from https://shorturl.at/OMmfX

USAID Measuring Impact Conservation Enterprise Retrospecti... | Flickr. (n.d.). Retrieved May 9, 2024, from https://www.flickr.com/photos/usaid-biodiversity-forestry/39396301085

CHAPTER XII

Case Studies and Real–World Applications

This chapter explores the transformative impact of bio-inspiration and engineering success stories through detailed case studies and real-world applications. It highlights how nature-inspired designs have led to significant innovations, such as Velcro from burrs, hydrodynamic surfaces from sharkskin, self-cleaning surfaces from lotus leaves, and passive cooling systems in buildings from termite mounds. Additionally, it delves into advanced technological developments inspired by biological phenomena, including gecko-inspired adhesives, bird wing-inspired aircraft designs, and whale fin-inspired wind turbine blades. The chapter also presents engineering triumphs like the Channel Tunnel, Burj Khalifa, Panama Canal expansion, NASA's Mars rovers, Tokyo Sky tree, Three Gorges Dam, and Japan's Shinkansen, illustrating the profoHeading 2und advancements across various engineering disciplines. These examples collectively demonstrate the power of nature-inspired innovation and engineering excellence in driving sustainable, efficient, and groundbreaking solutions across industries.

12.1 In-Depth Examples of Bio-Inspiration

Delving into the realm of bio-inspiration reveals a fascinating array of in-depth examples and real-world applications that underscore the transformative power of nature-inspired design.

A classic instance is the creation of Velcro, a ubiquitous fastening material, inspired by the tenacity with which burrs adhered to Swiss engineer George de Mestral's dog. Under a microscope, the hook-like structure of burrs led to the development of Velcro, featuring tiny hooks and loops for effective fastening. (Broeckhoven & Du Plessis, 2022; Hashemi Farzaneh & Lindemann, 2019)

Sharkskin, with its microscopic dermal denticles reducing water drag, has inspired surfaces applied in swimsuits and ship hulls, enhancing hydrodynamics and fuel efficiency.(Giorgio et al., 2021; Molina et al., 2020)

The Lotus Effect, observed in the water-repellent properties of lotus leaves, finds practical application in self-cleaning surfaces, from paints to architectural coatings, reducing the need for manual cleaning. (Broeckhoven & Du Plessis, 2022).

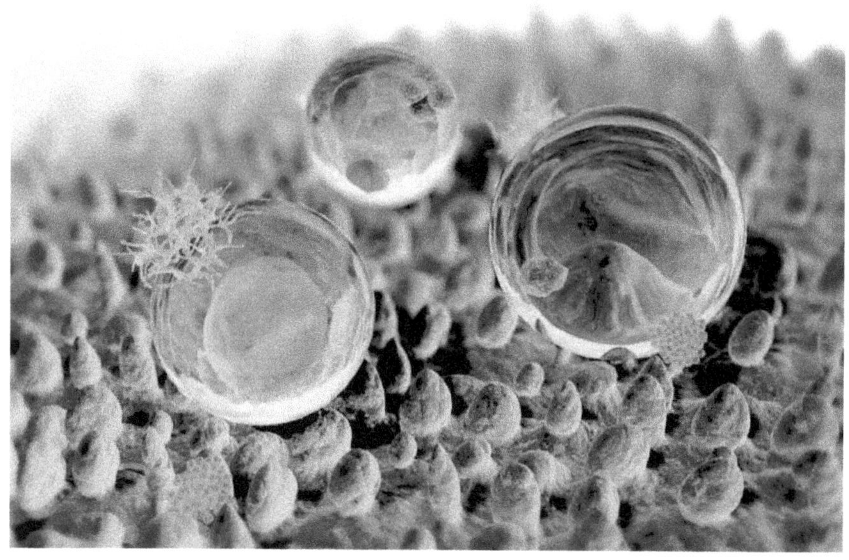

Fig.12.1: Lotus effect at microscopic level(*File:Lotus2mq.Jpg - Wikimedia Commons*, n.d.)

Sustainable building design takes inspiration from termite mounds, exemplified by Harare's Eastgate Centre, employing passive cooling systems akin to termite mounds' airflow regulation for energy-efficient ventilation.

Aircraft wing design draws from the efficiency of bird wings, with Airbus incorporating the "Sharklet" wingtip design to reduce drag, enhance fuel efficiency, and lower carbon emissions in the A320neo aircraft. Gecko-inspired adhesive technology mimics the microscopic hairs on a gecko's foot, facilitating strong and detachable adhesion used in robotics for climbing walls and navigating challenging terrains. Wind turbine blade design finds inspiration in whale fins, specifically the tubercles on humpback whale flippers that reduce drag and increase lift, leading to improved efficiency and reduced noise in whale-inspired wind turbine designs. (Broeckhoven & Du Plessis, 2022; Giorgio et al., 2021; Martinez et al., 2022, 2022)

Fig12.2: Humpback whale fin(*Two Humpback Whales Jumping out of the Water. Caudal Fin Humpback Whale Wal. - PICRYL - Public Domain Media Search Engine Public Domain Search*, n.d.)

In the realm of vision technology, engineers draw from insect compound eyes to develop bionic eyes featuring arrays of tiny lenses, applicable in surveillance, medical imaging, and robotics. These diverse examples vividly illustrate the translation of bio-inspiration into practical applications, showcasing its potential to drive sustainable and innovative solutions across industries.

12.2 Success Stories in Engineering

Exploring success stories in engineering reveals a spectrum of remarkable achievements through case studies and real-world applications.

The Channel Tunnel, linking the United Kingdom and France, exemplifies tunneling engineering excellence, stretching over 35 miles beneath the English Channel and employing innovative solutions like tunnel boring machines and precast concrete segments. (Martinez et al., 2022; Molina et al., 2020)

The Burj Khalifa in Dubai, the world's tallest building, completed in 2010, showcases architectural and structural engineering prowess, incorporating cutting-edge techniques such as a reinforced concrete structure and sophisticated wind load management. (Amer, 2019; Giorgio et al., 2021; Martinez et al., 2022)

The Panama Canal's expansion in 2016 stands as a testament to civil engineering achievements, enhancing global trade routes with the construction of new locks despite challenges in managing water resources, geological considerations, and massive concrete structures. (Blok & Gremmen, 2016; Budholiya et al., 2021)

NASA's Mars rover missions embody success in aerospace engineering, with rovers like Curiosity and Perseverance navigating Martian terrain and conducting experiments, demonstrating advancements in robotics, autonomous navigation, and engineering resilience in harsh environments. (Amer, 2019; Blok & Gremmen, 2016; Coyle et al., n.d.)

Tokyo Skytree, a telecommunications and observation tower in Japan completed in 2012, integrates seismic design, advanced materials, and cutting-edge elevator technology for stability and efficient vertical transportation. (Budholiya et al., 2021).

Fig.12.3: Flowers made from colourful papers(*Free Images: Plant, Flower, Petal, Balloon, Wind, Color, Colorful, Toy, Circle, Pinwheel, Cheerful, Art, Children, Illustration, Toys, Turn, Shape, Friendly, Screenshot, Windr Der 3264x2448 - - 1154653 - Free Stock Photos - PxHere, n.d.*)

The Three Gorges Dam on the Yangtze River in China, the world's largest power station completed in 2012, showcases success in hydroelectric engineering, addressing energy needs sustainably through intricate solutions for water flow management and flood prevention. (Coyle et al., n.d.; Giorgio et al., 2021)

Japan's Shinkansen, the Bullet Train, debuting in 1964, represents a triumph in transportation engineering, setting the standard for high-speed rail travel with innovations in aerodynamics, track design, and safety systems. (Martinez et al., 2022; Molina et al., 2020)

These success stories underscore the diverse contributions of engineering disciplines to infrastructure, transportation, space exploration, and sustainable energy, showcasing innovative thinking, technological advancements, and problem-solving capabilities defining the field of engineering.

Summary

This chapter provides an in-depth look at bio-inspiration and engineering success stories, showcasing how natural phenomena and innovative engineering solutions have led to significant advancements across various fields. The section on bio-inspiration highlights key examples such as Velcro, inspired by burrs; hydrodynamic surfaces modeled after sharkskin; self-cleaning technologies based on the Lotus Effect; termite mound-inspired passive cooling systems; bird wing-inspired aircraft designs; gecko-inspired adhesives; and whale fin-inspired wind turbine blades. These examples illustrate the practical applications and benefits of mimicking nature in design and technology.

The chapter also reviews notable engineering achievements, including the Channel Tunnel, the Burj Khalifa, the Panama Canal expansion, NASA's Mars rover missions, the Tokyo Sky tree, the Three Gorges Dam, and Japan's Shinkansen. These success stories exemplify the advancements and problem-solving capabilities in civil, structural, aerospace, and transportation engineering, highlighting the innovative thinking and technological progress that define modern engineering. Overall, the chapter underscores the importance of bio-inspiration and engineering ingenuity in creating sustainable, efficient, and groundbreaking solutions.

References

Amer, N. (2019). Biomimetic Approach in Architectural Education: Case study of 'Biomimicry in Architecture' Course. *Ain Shams Engineering Journal*.

Blok, V., & Gremmen, B. (2016). Ecological Innovation: Biomimicry as a New Way of Thinking and Acting Ecologically. *Journal of Agricultural and Environmental Ethics*, *29*(2), 203–217. https://doi.org/10.1007/s10806-015-9596-1

Broeckhoven, C., & Du Plessis, A. (2022). Escaping the Labyrinth of Bioinspiration: Biodiversity as Key to Successful Product Innovation. *Advanced Functional Materials*, *32*(18), 2110235. https://doi.org/10.1002/adfm.202110235

Budholiya, S., Bhat, A., Raj, S. A., Sultan, M. T. H., Shah, A. U., & Basri, A. A. (2021). *State of the Art Review about Bio-Inspired Design and Applications: An Aerospace Perspective*.

Coyle, S., Majidi, C., LeDuc, P., & Hsia, K. J. (n.d.). *Bio-inspired soft robotics: Material selection, actuation, and design*.

File:Lotus2mq.jpg—Wikimedia Commons. (n.d.). Retrieved May 15, 2024, from https://commons.wikimedia.org/wiki/File:Lotus2mq.jpg

Free Images: Plant, flower, petal, balloon, wind, color, colorful, toy, circle, pinwheel, cheerful, art, children, illustration, toys, turn, shape, friendly, screenshot, windr der 3264x2448—- 1154653—Free stock photos—PxHere. (n.d.). Retrieved May 15, 2024, from https://pxhere.com/en/photo/1154653#google_vignette

Giorgio, I., Spagnuolo, M., Andreaus, U., Scerrato, D., & Bersani, A. M. (2021). In-depth gaze at the astonishing mechanical behavior of bone: A review for designing bio-inspired hierarchical metamaterials. *Mathematics and Mechanics of Solids*, *26*(7), 1074–1103. https://doi.org/10.1177/1081286520978516

Hashemi Farzaneh, H., & Lindemann, U. (2019). *A Practical Guide to Bio-inspired Design*. Springer Berlin Heidelberg. https://doi.org/10.1007/978-3-662-57684-7

Martinez, A., Dejong, J., Akin, I., Aleali, A., Arson, C., Atkinson, J., Bandini, P., Baser, T., Borela, R., Boulanger, R., Burrall, M., Chen, Y., Collins, C., Cortes, D., Dai, S., DeJong, T., Del Dottore, E., Dorgan, K., Fragaszy, R., ... Zheng, J. (2022). Bio-inspired geotechnical engineering: Principles, current work, opportunities and challenges. *Géotechnique, 72*(8), 687–705. https://doi.org/10.1680/jgeot.20.P.170

Molina, D., Poyatos, J., Ser, J. D., García, S., Hussain, A., & Herrera, F. (2020). Comprehensive Taxonomies of Nature- and Bio-inspired Optimization: Inspiration Versus Algorithmic Behavior, Critical Analysis Recommendations. *Cognitive Computation, 12*(5), 897–939. https://doi.org/10.1007/s12559-020-09730-8

Two humpback whales jumping out of the water. Caudal fin humpback whale wal. - PICRYL - Public Domain Media Search Engine Public Domain Search. (n.d.). Retrieved May 15, 2024, from https://garystockbridge617.getarchive.net/media/caudal-fin-humpback-whale-wal-animals-b0993d

CHAPTER XIII

Challenges and Future Directions

This chapter addresses the ethical and environmental considerations and future prospects in bio-inspired engineering. Ethical concerns emphasize the importance of responsibly sourcing biological inspiration, avoiding harm to ecosystems, and maintaining ethical standards. Environmental considerations focus on minimizing the impact of biomimetic technologies through sustainable practices and life cycle assessments. The chapter also explores future directions, highlighting the complexity of biological systems, the need for interdisciplinary integration, advancements in robotics, the fusion of synthetic biology and bio-inspired engineering, and neuroengineering. Additionally, it discusses potential innovations in environmental monitoring, biologically augmented materials, and energy harvesting. The chapter underscores the importance of educational initiatives and ethical frameworks to guide the responsible development and deployment of bio-inspired technologies, offering a roadmap for transformative innovations and solutions to global challenges.

13.1 Ethical and Environmental Considerations

In the context of biomimicry, ethical and environmental considerations play pivotal roles in shaping the trajectory of this innovative field. Ethically, the exploration of nature-inspired solutions necessitates a conscientious approach to the sourcing of biological inspiration. Researchers and engineers must ensure that their endeavors adhere to ethical standards, avoiding harm to ecosystems and respecting the well-being of living organisms. This involves a delicate balance between drawing inspiration from nature's ingenious designs and ensuring responsible and sustainable practices in the development and application of biomimetic technologies. Environmental considerations, on the other hand, are intrinsic to the ethos of biomimicry. The very essence of this approach lies in emulating nature's efficiency and sustainability. As biomimetic solutions are developed and integrated into various industries, a commitment to minimizing environmental impact becomes imperative. This involves evaluating the entire life cycle of biomimetic technologies, from production to disposal, and striving to emulate nature's ability to contribute to ecological balance. Ultimately, the ethical and environmental dimensions of biomimicry underscore the importance of responsible innovation, urging practitioners to harness nature's brilliance with a profound sense of respect and accountability for the natural world. (Dicks, 2019; Mathews, 2011)

In the dynamic realm of biomimicry, ethical and environmental considerations emerge as pivotal determinants, significantly influencing the trajectory of this innovative field. Ethically, the exploration of solutions inspired by nature mandates a conscientious approach to the extraction of biological inspiration. Researchers and engineers bear the responsibility of ensuring that their pursuits align with ethical standards, steering clear of actions that could harm ecosystems and prioritizing the well-being of living organisms. Striking a delicate balance between drawing inspiration from nature's ingenious designs and upholding ethical standards becomes paramount, emphasizing the need for responsible and sustainable practices throughout the development and application of biomimetic technologies. (Dicks, 2017; Mathews, 2011, 2011).

Fig.13.1: Biomimetic 3D model of a flower(*File*, 2012)

Environmental considerations are inherently woven into the fabric of biomimicry. The essence of this approach lies in the emulation of nature's inherent efficiency and sustainability. As biomimetic solutions progress and find integration across diverse industries, a steadfast commitment to minimizing environmental impact becomes imperative. This commitment encompasses a comprehensive assessment of the entire life cycle of biomimetic technologies, spanning from their production to eventual disposal. Aligning with nature's ability to contribute to ecological balance, practitioners of biomimicry strive not only to mimic nature's brilliance but also to embody its intrinsic environmental stewardship. The ethical and environmental dimensions of biomimicry converge to underscore the paramount importance of responsible innovation. Practitioners are called upon to harness the brilliance of nature with not only technical proficiency but also with a profound sense of respect and accountability for the intricate and interconnected web of the natural world. In essence, the ethical and environmental considerations within biomimicry embody a commitment to harmonize human innovation with the delicate balance of nature. (Bensaude-Vincent, n.d.; Dicks, 2019)

13.2 Future Prospects and Unexplored Territories in Bio-Inspired Engineering

The challenges and future directions in bio-inspired engineering delineate a landscape rich with potential and unexplored territories. As the field continues to evolve, several key aspects shape its future prospects. One prominent challenge lies in unraveling the complexity of biological systems. While bio-inspired engineering draws inspiration from nature's intricacies, understanding and replicating these complex systems pose ongoing challenges that researchers are poised to tackle.

Interdisciplinary integration stands out as a future direction, emphasizing collaboration across diverse scientific domains. The fusion of biology, engineering, chemistry, and other fields fosters a holistic approach to innovation, pushing the boundaries of what bio-inspired engineering can achieve. (Ersanlı & Ersanlı, n.d.; Hwang et al., n.d.)

Advancements in robotics, including more sophisticated and adaptable designs inspired by animal locomotion and swarm intelligence, represent an unexplored territory. The future holds the promise of robots with enhanced agility, autonomy, and versatility, mirroring the capabilities observed in the natural world.(Coyle et al., n.d.; González-Aguirre et al., 2021)

The integration of synthetic biology with bio-inspired engineering introduces another exciting dimension. Biohybrid systems, combining living and synthetic components, hold unexplored potential for enhanced functionalities, ranging from bio-hybrid robots to biocompatible electronics.

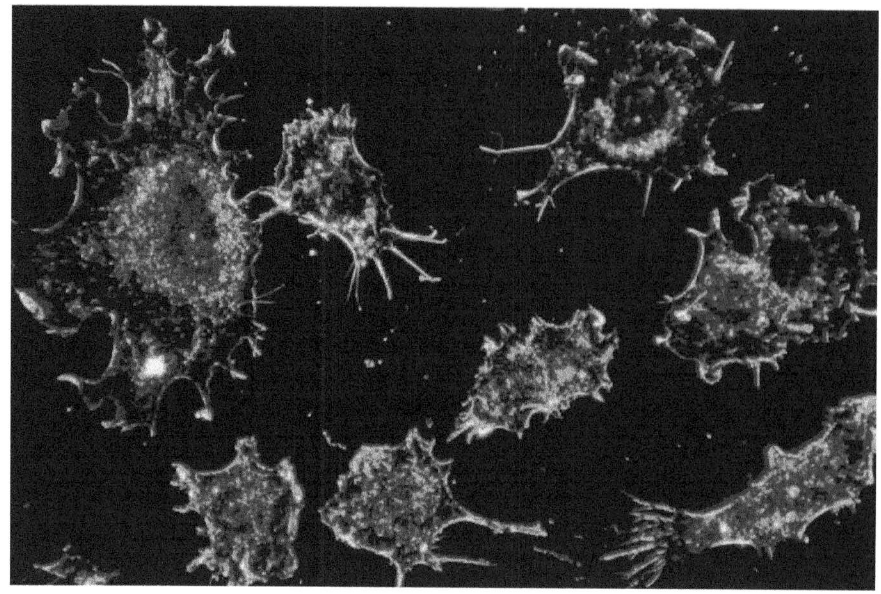

Fig.13.2: An AI image depicting connections between neurons(Staff, 2022)

Neuro engineering emerges as a future prospect, aiming to mimic and understand the complexities of the human brain. This could lead to breakthroughs in brain-machine interfaces, neuro prosthetics, and cognitive computing, pushing the boundaries of what is achievable in artificial intelligence and human-machine interaction. (*Human Brain Neurons: Understanding the Intricacies of Neural Networks | AI Art Generator | Easy-Peasy.AI*, n.d.)

In the realm of environmental monitoring and remediation, unexplored territories involve developing bio-inspired technologies for pollution detection, air and water purification, and sustainable environmental management. Mimicking natural systems, such as plants or microorganisms, could pave the way for innovative solutions to pressing environmental challenges.(Dicks, 2019)

Biologically augmented materials, drawing inspiration from the structures and properties of living organisms, offer uncharted territories in material science. Future materials may include lightweight and strong composites, self-healing polymers, and adaptive surfaces with applications spanning aerospace, construction, and consumer goods.(Iouguina et al., 2014)

Energy harvesting and storage solutions inspired by plants present an unexplored frontier. Learning from nature's efficiency in energy conversion and storage, engineers have the potential to develop advanced technologies for sustainable energy production and storage.(Chayaamor-Heil & Hannachi-Belkadi, 2017)

As bio-inspired engineering progresses, educational initiatives and ethical considerations are gaining prominence. Unexplored territories involve creating comprehensive educational programs to increase awareness and understanding of bio-inspired engineering. Ethical frameworks and guidelines for responsible and sustainable practices in the development and deployment of bio-inspired technologies are crucial aspects that will shape the future trajectory of the field.(Hwang et al., n.d.).

Fig 13.3: A model made by using biological inspiration(*Mixing in Biology, Inspired by the Fantastic Book "art For... | Flickr*, n.d.)

In essence, the challenges and future directions in bio-inspired engineering offer a roadmap towards uncharted territories, where the intersection of biology and engineering holds the promise of transformative innovations and solutions to complex global challenges.

Summary

This chapter explores the ethical and environmental considerations as well as future prospects and challenges in bio-inspired engineering. The ethical dimension emphasizes the need for responsible sourcing of biological inspiration, avoiding harm to ecosystems, and adhering to high ethical standards. Environmental considerations focus on the sustainability of biomimetic technologies, aiming to minimize their impact throughout their life cycle.

The chapter also outlines future directions in bio-inspired engineering, highlighting several key areas. These include the complexity of understanding and replicating biological systems, the importance of interdisciplinary collaboration, advancements in robotics inspired by animal locomotion and swarm intelligence, and the integration of synthetic biology with bio-inspired engineering. Other promising areas are neuro engineering, which aims to mimic the human brain, and the development of bio-inspired technologies for environmental monitoring and remediation.

Further, the chapter discusses the potential of biologically augmented materials, such as self-healing polymers and adaptive surfaces, and energy solutions inspired by plants. It also stresses the importance of developing educational programs and ethical guidelines to ensure responsible innovation. Overall, Chapter 13 provides a comprehensive overview of the current challenges and future directions in bio-inspired engineering, emphasizing the potential for transformative innovations that address complex global challenges.

References

Bensaude-Vincent, B. (n.d.). Bio-Informed Emerging Technologies and Their Relation to the Sustainability Aims of Biomimicry. *Environmental Values.*

Chayaamor-Heil, N., & Hannachi-Belkadi, N. (2017). *Towards a Platform of Investigative Tools for Biomimicry as a New Approach for Energy-Efficient Building Design.*

Coyle, S., Majidi, C., LeDuc, P., & Hsia, K. J. (n.d.). *Bio-inspired soft robotics: Material selection, actuation, and design.*

Dicks. (2019). The Biomimicry Revolution in Environmental Epistemology. *Ethics and the Environment, 24*(2), 43. https://doi.org/10.2979/ethicsenviro.24.2.03

Dicks, H. (2017). The Poetics of Biomimicry. *Environmental Philosophy, 14*(2), 191–219. https://www.jstor.org/stable/26894352

Ersanlı, E. T., & Ersanlı, C. C. (n.d.). Biomimicry: Journey to the Future with the Power of Nature. *International Scientific and Vocational Studies Journal, 7*(2), 149–160. Retrieved May 15, 2024, from https://dergipark.org.tr/en/pub/bilmes/issue/82358/1388402

File:Biomimetic Morphology of Phyllotaxy Towers.jpg—Wikimedia Commons. (2012, October 11). https://commons.wikimedia.org/wiki/File:Biomimetic_Morphology_of_Phyllotaxy_Towers.jpg

González-Aguirre, J. Á., Osorio-Oliveros, R., Rodríguez-Hernández, K. L., Lizárraga-Iturralde, J., Menendez, R. M., Ramírez-Mendoza, R. A., & Ramírez-Moreno, M. A. (2021). *Service Robots: Trends and Technology.*

Human Brain Neurons: Understanding the Intricacies of Neural Networks | AI Art Generator | Easy-Peasy.AI. (n.d.). Retrieved May 9, 2024, from https://easy-peasy.ai/ai-image-generator/images/human-brain-neurons-complex-network-dendrites-axons

Hwang, J., Jeong, Y., Park, J. M., Lee, K. H., Wook, J., & Choi, J. (n.d.). Biomimetics: Forecasting the future of science, engineering, and medicine. *International Journal of Nanomedicine.*

Iouguina, A., Dawson, J. W., Hallgrimsson, B., & Smart, G. (2014). *BIOLOGICALLY INFORMED DISCIPLINES: A COMPARATIVE ANALYSIS OF BIONICS, BIOMIMETICS, BIOMIMICRY, AND BIO-INSPIRATION AMONG OTHERS. 9*(3).

Mathews, F. (2011). Towards a Deeper Philosophy of Biomimicry. *Organization & Environment, 24*(4), 364–387. https://doi.org/10.1177/1086026611425689

Mixing in biology, inspired by the fantastic book "art for... | Flickr. (n.d.). Retrieved May 15, 2024, from https://www.flickr.com/photos/_sk/23087005352

Staff, N. N. (2022, November 2). NCSA Project Expands Toolbox for Understanding Cancer Evolution. *NCSA*. https://www.ncsa.illinois.edu/ncsa-project-expands-toolbox-for-understanding-cancer-evolution/

CHAPTER XIV

Conclusion and Outlook

This chapter provides a comprehensive conclusion and outlook on the field of biomimicry, summarizing key takeaways and future prospects. It highlights the convergence of biology and engineering, emphasizing nature as a source of innovative solutions honed through evolution. Interdisciplinary collaboration, ethical and environmental considerations, and the dynamic nature of biomimicry are identified as critical components of this field. The chapter also explores future prospects, including advancements in robotics, synthetic biology, neuro engineering, and environmental applications. These prospects reflect the potential for bio-inspired engineering to revolutionize various industries and contribute to sustainable and responsible innovation. Ultimately, the chapter underscores the transformative power of biomimicry, advocating for continued exploration and appreciation of nature's ingenious designs.

14.1 Summarizing Key Takeaways

In the realm of biomimicry, the conclusion and outlook underscore key takeaways that encapsulate the essence of this innovative discipline. At its core, biomimicry is a convergence of biology and engineering, drawing inspiration from nature's time-tested designs to address contemporary challenges. One key takeaway is the recognition of nature as a prolific innovator, offering a vast repository of solutions honed through millions of years of evolution. Biomimicry, therefore, serves as a conduit for translating these natural solutions into human-made technologies, fostering sustainable and efficient designs.(Amer, 2019; Bensaude-Vincent, n.d.)

The significance of interdisciplinary collaboration emerges as a pivotal takeaway. Biomimicry necessitates the integration of expertise from diverse fields, including biology, engineering, materials science, and more. This collaborative approach not only mirrors the interconnectedness of ecosystems but also enhances the depth and breadth of innovation, allowing for holistic solutions inspired by nature.(Fu et al., 2014).

Fig.14.1: A building covered with green trees(*Royalty-Free Photo: House Surrounded by Green Leafed House | PickPik*, n.d.)

Ethical and environmental considerations stand out as crucial takeaways in the context of biomimicry. Practitioners in this field must navigate the delicate balance between extracting biological inspiration responsibly and ensuring that biomimetic technologies contribute to environmental stewardship. Acknowledging the ethical dimensions underscores the importance of upholding ethical standards and safeguarding ecosystems during the pursuit of bio-inspired solutions. (Blok & Gremmen, 2016; Speck et al., 2017).

The conclusion and outlook also emphasize the dynamic nature of biomimicry, pointing towards unexplored territories and future prospects. Challenges such as unraveling the complexity of biological systems and advancing interdisciplinary collaborations represent ongoing focal points. The outlook anticipates advancements in robotics, synthetic biology integration, neuro engineering and environmental applications, reflecting the ever-expanding horizons of biomimicry. (Budholiya et al., 2021)

Ultimately, biomimicry offers a transformative approach to innovation, where nature serves as both a mentor and a model. The key takeaways encapsulate the foundational principles of respecting and learning from the natural world, fostering collaboration across disciplines, and embracing a future where bio-inspired solutions contribute to a sustainable and harmonious coexistence with our planet. As biomimicry continues to evolve, it holds the promise of unlocking novel solutions to intricate challenges while instilling a profound appreciation for the ingenious designs found in the diversity of life on Earth. (Broeckhoven & Du Plessis, 2022).

Fig.14.2: Flowers made from colourful papers(*Free Images : Plant, Flower, Petal, Balloon, Wind, Color, Colorful, Toy, Circle, Pinwheel, Cheerful, Art, Children, Illustration, Toys, Turn, Shape, Friendly, Screenshot, Windr Der 3264x2448 - - 1154653 - Free Stock Photos - PxHere*, n.d.)

14.2 Future Prospects for Bio-Inspired Engineering

In concluding the exploration of bio-inspired engineering, the future prospects unfold as a tapestry of innovation and transformative potential. Bio-inspired engineering stands at the intersection of biology and technology, offering a glimpse into a future where nature's ingenious solutions guide human innovation. One prominent future prospect lies in the continued advancement of robotics, where designs inspired by the efficiency and adaptability of natural organisms promise to redefine the capabilities of autonomous systems. This could lead to the development of more agile, versatile, and intelligent robots, revolutionizing industries from healthcare to manufacturing. (Ersanlı & Ersanlı, n.d.)

Synthetic biology integration presents another compelling avenue for future exploration. The convergence of synthetic biology with bio-inspired engineering opens doors to the creation of biohybrid systems, blending living and synthetic components. This promises breakthroughs in fields ranging from medicine, where biohybrid implants could enhance human health, to environmental applications, where engineered organisms contribute to sustainable solutions. (Speck et al., 2017)

Neuro engineering emerges as a frontier with significant promise. As our understanding of the intricate workings of the human brain deepens, bio-inspired approaches may lead to innovations in brain-machine interfaces, neuro prosthetics, and cognitive computing. This could pave the way for transformative advancements in healthcare, communication, and human-computer interaction. (Amer, 2019)

Environmental applications of bio-inspired engineering are poised to play a pivotal role in addressing pressing global challenges. From sustainable materials inspired by nature to technologies for environmental monitoring and remediation, bio-inspired solutions offer a pathway to a more ecologically conscious future. The development of biologically augmented materials, energy harvesting inspired by plants, and eco-friendly design principles holds the potential to reshape industries with a focus on sustainability. (Amer, 2019; Bensaude-Vincent, n.d.)

Education and ethical considerations are integral to the future of bio-inspired engineering. Increased awareness and understanding of biomimicry, coupled with robust educational initiatives, will foster a new generation of innovators equipped to tackle complex challenges through nature-inspired solutions. Ethical frameworks and guidelines will continue to evolve, ensuring responsible practices in the development and deployment of bio-inspired technologies. (Amer, 2019)

In essence, the future prospects for bio-inspired engineering paint a picture of continuous discovery and innovation. The field is poised to contribute not only to technological advancements but also to a paradigm shift in how we approach design, sustainability, and the intricate interconnectedness of the natural world. As bio-inspired engineering unfolds in the years to come, it holds the promise of unlocking transformative solutions that harmonize human progress with the brilliance of the natural world.

Summary

This chapter concludes the exploration of biomimicry by summarizing key takeaways and outlining future prospects for bio-inspired engineering. It underscores the core principle of biomimicry: drawing inspiration from nature's time-tested designs to create sustainable and efficient technologies. The chapter highlights the importance of interdisciplinary collaboration, ethical practices, and environmental stewardship in advancing biomimetic innovations.

It also discusses future directions, emphasizing potential breakthroughs in robotics, synthetic biology, neuro engineering, and environmental applications. These advancements promise to revolutionize various fields, from healthcare and manufacturing to ecological management. The chapter calls for robust educational initiatives and ethical frameworks to guide the responsible development and application of bio-.Overall, this chapter paints an optimistic outlook, showcasing biomimicry's potential to harmonize human progress with nature's brilliance and contribute to sustainable, innovative solutions for complex global challenges inspired technologies

.

References

Amer, N. (2019). Biomimetic Approach in Architectural Education: Case study of 'Biomimicry in Architecture' Course. *Ain Shams Engineering Journal*.

Bensaude-Vincent, B. (n.d.). Bio-Informed Emerging Technologies and Their Relation to the Sustainability Aims of Biomimicry. *Environmental Values*.

Blok, V., & Gremmen, B. (2016). Ecological Innovation: Biomimicry as a New Way of Thinking and Acting Ecologically. *Journal of Agricultural and Environmental Ethics, 29*(2), 203–217. https://doi.org/10.1007/s10806-015-9596-1

Broeckhoven, C., & Du Plessis, A. (2022). Escaping the Labyrinth of Bioinspiration: Biodiversity as Key to Successful Product Innovation. *Advanced Functional Materials, 32*(18), 2110235. https://doi.org/10.1002/adfm.202110235

Budholiya, S., Bhat, A., Raj, S. A., Sultan, M. T. H., Shah, A. U., & Basri, A. A. (2021). *State of the Art Review about Bio-Inspired Design and Applications: An Aerospace Perspective.*

Ersanlı, E. T., & Ersanlı, C. C. (n.d.). Biomimicry: Journey to the Future with the Power of Nature. *International Scientific and Vocational Studies Journal, 7*(2), 149–160. Retrieved May 15, 2024, from https://dergipark.org.tr/en/pub/bilmes/issue/82358/1388402

Free Images: Plant, flower, petal, balloon, wind, color, colorful, toy, circle, pinwheel, cheerful, art, children, illustration, toys, turn, shape, friendly, screenshot, windr der 3264x2448—- 1154653—Free stock photos—PxHere. (n.d.). Retrieved May 15, 2024, from https://pxhere.com/en/photo/1154653

Fu, K., Moreno, D., Yang, M., & Wood, K. L. (2014). Bio-inspired design: An overview investigating open questions from the broader field of design-by-analogy. *Journal of Mechanical Design, 136*(11), 111102. https://asmedigitalcollection.asme.org/mechanicaldesign/article-abstract/136/11/111102/375236

Royalty-Free photo: House surrounded by green leafed house | PickPik. (n.d.). Retrieved May 15, 2024, from https://www.pickpik.com/house-creeper-ivy-green-wales-architecture-154210

Speck, O., Speck, D., Horn, R., Gantner, J., & Sedlbauer, K. P. (2017). Biomimetic bio-inspired biomorph sustainable? An attempt to classify and clarify biology-derived technical developments. *Bioinspiration & Biomimetics, 12*(1), 011004. https://iopscience.iop.org/article/10.1088/1748-3190/12/1/011004/meta

CHAPTER XV

How Methodology works in Biomimicry

This chapter delves into the methodology of biomimicry, outlining a systematic approach to solving human challenges by drawing inspiration from nature. The process involves several key steps: defining the problem, conducting biological research, finding analogous inspirations, abstracting principles, applying these principles to human designs, and iterative testing and refinement. This methodology aims to leverage nature's time-tested solutions to create sustainable and efficient technologies.

The chapter also explores future applications of biomimicry that do not yet exist but hold potential. These include:

1. Energy Harvesting Inspired by Plants: Improving solar energy technologies by mimicking photosynthesis.

2. Adaptive Camouflage Inspired by Octopuses: Developing materials capable of dynamic color and texture changes for enhanced concealment.

3. Structural Materials Inspired by Spider Silk: Creating stronger and more flexible materials by replicating the molecular structure of spider silk.

4. Communication Systems Inspired by Bees: Optimizing communication networks using principles from bee colony communication.

5. Self-healing Materials Inspired by Skin: Developing materials that can autonomously repair damage.

6. Efficient Water Collection Inspired by Desert Beetles: Designing surfaces that mimic the water-harvesting abilities of the Namib Desert beetle for use in arid environments.

By following these methodologies, biomimicry aims to address current limitations in various fields, offering innovative and sustainable solutions inspired by nature's efficiency and adaptability. In biomimicry, methodology involves observing and imitating nature's solutions to solve human challenges. It typically follows these steps:

STEP 1. Define the Problem: Clearly identify the problem or challenge you want to address.

STEP 2. Biological Research: Study and analyze biological organisms and systems relevant to the problem. Understand how nature has already solved similar issues.

STEP 3. Analogous Inspiration: Identify organisms, processes, or systems in nature that offer solutions or insights applicable to the human problem.

STEP 4. Abstraction: Extract key principles or strategies from the biological inspiration that can be applied to human design or technology.

STEP 5. Application: Integrate the extracted principles into human design or technology to develop innovative solutions.

STEP 6. Testing and Refinement: Test the biomimetic design, gather feedback, and refine the solution based on performance and functionality.

STEP 7. Iterate: Repeat the process if necessary, refining the design through multiple iterations.

By mimicking nature's time-tested strategies, biomimicry aims to create sustainable and efficient solutions across various fields, from engineering to product design.

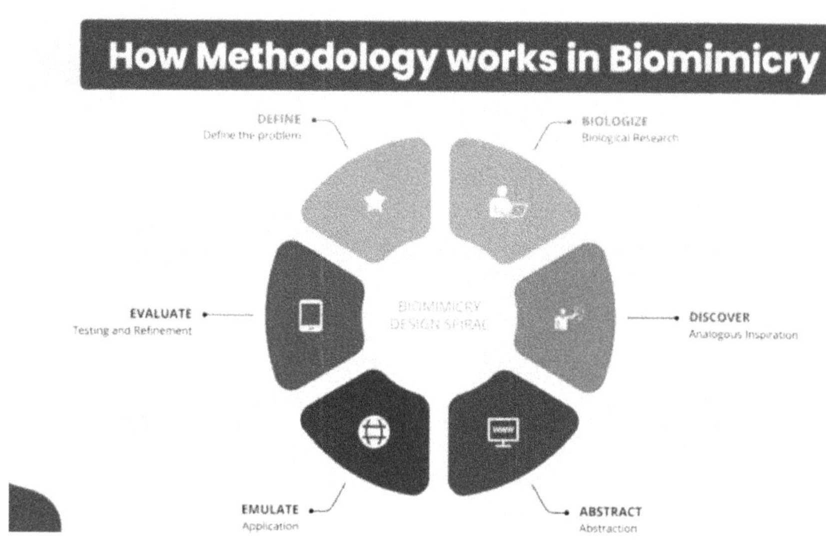

Examples of Biomimicry that doesn't exist today but will have scope for future existence

EXAMPLE 1. Energy Harvesting Inspired by Plants: Developments in mimicking the efficient energy conversion process of photosynthesis in plants could lead to improved solar energy harvesting technologies. (Araque et al., 2021; Krishna & Tiwari, 2021, 2021; Z. Wang et al., 2022).

STEP 1: DEFINE THE PROBLEM

Problems related to energy harvesting that could find solutions through inspiration from plants include:

Solar energy conversion faces challenges such as inefficiency in traditional panels, which struggle to capture sunlight effectively. Biomimicry offers a solution by emulating plant photosynthesis to enhance conversion rates. Additionally, renewable sources like solar and wind encounter intermittency issues. Mimicking how plants store and release energy may lead to improved storage solutions, reducing reliance on continuous sunlight or wind. Moreover, conventional solar panels struggle in low-light conditions, but biomimetic approaches inspired by plants could enhance energy harvesting capabilities in varying light conditions as traditional solar panel manufacturing is resource-intensive, prompting exploration of biomimicry for sustainable alternatives. By emulating efficient energy conversion processes in plants, biomimetic approaches may use organic or bio-inspired materials. Additionally, addressing the impact of land use on ecosystems, biomimicry can aid in developing energy-harvesting systems that mimic plants' efficient space utilization for photosynthesis. Researchers aspire to leverage plant-inspired solutions to create sustainable and efficient energy harvesting methods, thereby contributing to the shift towards cleaner and renewable energy sources. . (Araque et al., 2021; Krishna & Tiwari, 2021, 2021; Z. Wang et al., 2022).

Fig15.1: A solar pannel(*Solar Panel - Google Search*, n.d.)

STEP 2: BIOLOGICAL RESEARCH

Biological research focuses on comprehending the mechanisms and intricacies of photosynthesis—the fundamental energy conversion process in plants. This investigation encompasses the examination of photosynthetic structures such as chloroplasts and thylakoid membranes, seeking to unravel how these components capture and convert sunlight into chemical energy. Pigment analysis, particularly of chlorophyll and carotenoids, plays a crucial role in understanding how plants optimize light absorption for efficient energy conversion across different wavelengths. Additionally, researchers delve into the intricate pathways facilitating energy transfer within cellular structures, ultimately leading to the production of adenosine triphosphate (ATP) and other energy-rich molecules. . (Araque et al., 2021; Krishna & Tiwari, 2021, 2021; Z. Wang et al., 2022).

Fig.15.2: A sunflower feild

The study extends to how plants adapt to diverse light conditions, including low-light and fluctuating environments, with a focus on molecular and cellular responses. Investigating storage mechanisms during excess energy production further informs the development of energy storage solutions inspired by natural processes. Biomolecular processes, involving proteins, enzymes, and other molecular components, are scrutinized to identify potential bio-inspired materials for energy harvesting. This comprehensive exploration of photosynthesis and plant adaptations provides valuable insights guiding the design of innovative energy harvesting systems inspired by the remarkable efficiency of nature.

STEP 3 : ANALOGOUS INSPIRATION

Analogous involves identifying and drawing parallels between the biological processes observed in plants, particularly photosynthesis, and the requirements or challenges in human-designed energy harvesting systems. It entails recognizing how nature has efficiently captured and converted solar energy and translating those principles into design concepts for more effective and sustainable energy technologies. This could include mimicking the structures, mechanisms, and adaptive strategies found in plants to enhance the efficiency , reliability, and versatility of energy harvesting solutions. . (Araque et al., 2021; Krishna & Tiwari, 2021, 2021; Z. Wang et al., 2022).

Fig.15.3: Solar pannels on the roof of house(*What Is Solar Panel Efficiency - Prime Energy Solar*, n.d.)

STEP 4 : ABSTRACTION:

Abstraction refers to distilling the fundamental principles and key features identified during the study of photosynthesis and plant energy processes. It involves extracting the essential elements from biological research and analogous inspiration to form a generalized understanding of how plants efficiently capture, convert, and store energy from sunlight. This abstraction guides the application of these principles to design human-made energy harvesting systems, focusing on the core concepts rather than replicating every intricate detail of the biological processes.

STEP 5 : APPLICATION:

Application involves implementing the abstracted principles and insights gained from biological research into the design and development of human-made energy harvesting systems. This phase focuses on translating the biomimetic concepts into practical solutions for improved energy capture, conversion, and storage. It encompasses the integration of bio-inspired designs, materials, or processes into technologies that mimic nature's efficiency in harnessing solar energy. The goal is to create innovative and sustainable energy harvesting applications based on the lessons learned from the biological world. (*Money Matters*, n.d.).

STEP 6 : TESTING AND REFINEMENT

The testing and refinement process follow a systematic approach:

Commencing with prototype testing, biomimetic designs are translated into prototypes, and their performance in capturing and converting energy is rigorously examined, considering factors like efficiency, reliability, and adaptability to diverse conditions. The evaluation extends to comparing the energy conversion efficiency of the biomimetic system with traditional technologies under various environmental scenarios. The system's adaptability to different light conditions is scrutinized, mirroring the observed adaptability in plants. Durability and longevity assessments are conducted, considering wear and tear, weather resistance, and the system's ability to sustain performance over time. If the biomimetic design incorporates energy storage inspired by plant mechanisms, the effectiveness of these storage solutions is thoroughly tested, including the system's capacity to store and release energy as needed. Environmental impact considerations examine the system's alignment with sustainability goals and its ecological footprint compared to conventional technologies. User feedback becomes integral, with stakeholders providing insights for refinement. This feedback-driven optimization process informs an iterative design approach, where findings from testing are applied to enhance overall performance, addressing any identified challenges or limitations.

Through this testing and refinement process, the biomimetic energy harvesting system can be fine-tuned to achieve optimal performance, reliability, and sustainability.

EXAMPLE 2. Adaptive Camouflage Inspired by Octopuses: Future materials might be designed to mimic the adaptive camouflage abilities of octopuses, allowing for dynamic color and texture changes in response to the environment. (Boghossian et al., 2011; Bu & Bai, 2023; Han et al., 2021; Hanlon et al., 2009, 2009; K. Li et al., 2022; Y. Li et al., 2018; Nakajima et al., 2022; Wilson et al., 2021).

Fig.15.4: An image of octopus(*Mollusc,Octopoda,Eight,Arms - Free Image from Needpix.Com*, n.d.)

STEP 1: DEFINE THE PROBLEM

The problem is the limited adaptability of current camouflage materials, which struggle to replicate the dynamic color and texture changes observed in octopuses. Traditional camouflage is often static and lacks the ability to adjust in real-time to different environmental conditions. The goal is to overcome this limitation and develop materials that can mimic the adaptive camouflage abilities of octopuses, offering enhanced concealment and versatility in various settings.

STEP 2: BIOLOGICAL RESEARCH

To tackle the challenge at hand, extensive biological research on octopuses is imperative to unravel the intricacies of their adaptive camouflage. This research encompasses a multi-faceted approach:

Firstly, a detailed examination of octopus skin structure delves into the arrangement of pigment cells, specifically chromatophores, and their role in facilitating rapid changes in both color and texture. Neural control mechanisms governing the activation of chromatophores in response to visual stimuli are scrutinized, alongside an exploration of how octopuses process environmental cues to initiate camouflage adjustments.

Observing octopuses in their natural habitats is a crucial aspect, shedding light on their interactions with diverse environments. This includes analyzing the spectrum of color and texture changes in varying surroundings and potential threat scenarios. The investigation extends to understanding the adaptive strategies employed by octopuses for camouflage, encompassing considerations such as predation, mating, and communication. This exploration also entails scrutinizing how octopuses adeptly blend into backgrounds and employ disruptive patterns for concealment.

A focus on material properties involves examining octopus skin's physical attributes, emphasizing flexibility, durability, and light-reflecting capabilities. Comprehending how these properties contribute to effective adaptive camouflage forms a pivotal aspect of the research. Additionally, the impact of environmental factors, ranging from light intensity and temperature to substrate, on octopus camouflage is thoroughly assessed. This encompasses identifying the sensory cues that prompt adaptive responses in camouflage.

The overarching goal of this comprehensive biological research is to garner profound insights into the intricate mechanisms governing octopus adaptive camouflage. Such knowledge forms the bedrock for the development of innovative materials capable of replicating these remarkable capabilities, enabling dynamic color and texture changes in response to diverse environments.

STEP 3: ANALOGOUS INSPIRATION

Identifying analogous systems or processes for translation into material design for adaptive camouflage involves several key considerations:

Firstly, the exploration of materials or technologies that mimic the functionality of chromatophores, such as color-changing pigments or microstructures that can be controlled to alter appearance. This avenue seeks to replicate the dynamic color control observed in octopuses.

Secondly, the investigation delves into artificial neural networks or algorithms capable of replicating the neural control mechanisms observed in octopuses. This exploration aims to facilitate real-time responses to environmental stimuli, mirroring the adaptive neural control seen in the natural counterparts.

Additionally, the quest involves searching for materials with properties akin to octopus skin, prioritizing flexibility, durability, and the capacity to reflect and manipulate light. This endeavor aims to reproduce the physical attributes contributing to effective adaptive camouflage observed in octopuses.

Lastly, the identification of technologies capable of replicating the dynamic texture changes observed in octopuses is essential. This involves considering materials with the ability to alter surface patterns and textures, thus emulating the versatile texture adaptation observed in octopus camouflage.

STEP 4: ABSTRACTION

Abstraction is a crucial step involving the distillation of identified analogous inspirations into fundamental principles and design elements, facilitating the extraction of essential features for integration into the development of adaptive camouflage materials. This process involves several key abstractions:

Firstly, the abstraction of dynamic color control entails focusing on the underlying mechanisms of chromatophores and translating them into controllable material features. This abstraction seeks to capture the ability to dynamically control color changes, a fundamental aspect observed in octopuses.

Secondly, the abstraction of responsive neural algorithms involves translating the neural control aspects observed in octopuses into algorithms or control systems capable of interpreting environmental cues and initiating adaptive responses in materials. This abstraction aims to replicate the adaptive neural control observed in natural counterparts.

Additionally, the abstraction process extends to mimicking skin flexibility, distilling the flexibility and texture-changing capabilities of octopus skin into material properties that can be replicated in synthetic materials. This abstraction is crucial for achieving the desired adaptability and versatility in the adaptive camouflage materials.

Lastly, the abstraction of incorporating light reflection involves considering how to mimic the light-reflecting properties of octopus skin in adaptive materials for effective concealment. This abstraction aims to capture the optical features contributing to the success of octopus camouflage in diverse environments.

STEP 5: APPLICATION

The application step involves the seamless integration of the abstracted principles into the practical development of adaptive camouflage materials. This comprehensive application encompasses various considerations:

Initially, material development takes center stage, where advanced materials are harnessed, incorporating color-changing pigments, flexible structures, and responsive elements inspired by the identified analogous systems. This step forms the foundational basis for creating adaptive camouflage materials with enhanced capabilities.

Next, the integration of sensors or environmental monitoring systems becomes crucial. These sensors feed real-time information to the material, enabling dynamic adjustments based on surrounding conditions. This integration is pivotal for ensuring the adaptability and responsiveness of the adaptive camouflage materials.

Moving forward, the development process advances to prototyping and testing. Prototypes of adaptive camouflage materials are meticulously crafted, and thorough testing is conducted to assess their effectiveness in diverse environments and scenarios. This step ensures the functionality and practical applicability of the materials under varying conditions.

Lastly, a strategic consideration involves identifying potential industries and applications where adaptive camouflage materials could be transformative. Industries such as military, wildlife conservation, and surveillance emerge as key areas where these innovative materials can make a significant impact.

EXAMPLE 3. Structural Materials Inspired by Spider Silk: Engineers might create stronger and more flexible materials by replicating the molecular structure of spider silk, which is known for its exceptional strength and elasticity.(Dumanli & Savin, 2016; Humenik et al., 2011; Koh et al., 2015; Lefèvre & Auger, 2016a, 2016b; Su et al., 2020; Y. Wang et al., 2018, 2020; N. Zhao et al., 2014; Zhou et al., 2018).

Fig.15.5: Spider web(*Spider Web on Abstract Blur Green Background - PatternPictures*, n.d.)

Step 1: Define the Problem in the Context of Biomimicry:
Recognize the limitations of current structural materials in achieving optimal strength and flexibility. Acknowledge that nature, specifically spider silk, exhibits exceptional properties in these aspects. The problem lies in the inability of conventional materials to match the unique combination of strength and elasticity found in spider silk. The goal is to redefine material design by formulating a problem statement that challenges engineers to replicate the molecular structure of spider silk, thereby creating materials with superior strength and flexibility through biomimicry.

<u>Step 2: Biological Research:</u>
In the methodology of biomimicry, biological research involves a deep exploration of the natural world, specifically focusing on organisms or systems that exhibit desired characteristics. In the case of structural materials inspired by spider silk, this step entails studying the molecular structure, production process, and properties of spider silk through biological research. By understanding the intricacies of spider silk at a

biological level, engineers gain insights into how nature achieves the remarkable combination of strength and elasticity.

<u>Step 3: Analogous Inspiration</u>

Following biological research, the next step is to draw analogous inspiration. This involves identifying and translating the key principles observed in spider silk into engineering and design concepts. Engineers seek to apply the lessons learned from nature to develop analogous solutions for creating stronger and more flexible materials. This step bridges the gap between biological insights and their potential application in engineering.

<u>Step 4: Abstraction</u>

Abstraction is the process of distilling essential principles from the biological model. In the context of spider silk-inspired materials, it entails extracting the fundamental design concepts responsible for the exceptional strength and elasticity of spider silk. By abstracting these principles, engineers create a set of guidelines and rules that can be applied in the design of synthetic materials while retaining the core attributes observed in nature.

<u>Step 5: Application</u>

The final step involves applying the abstracted principles to the development and manufacturing of new materials. Engineers utilize the biomimetic approach to replicate the molecular structure of spider silk, incorporating the design concepts derived from nature. This application phase aims to produce real-world solutions—materials that exhibit enhanced strength and flexibility, mimicking the success of spider silk. The iterative process of applying, testing, and refining ensures that the biomimetic design achieves its intended goals.

EXAMPLE 4. Communication Systems Inspired by Bees: Future communication networks could be optimized by drawing inspiration from the efficient communication systems observed in bee colonies.(Aghakhani et al., 2023; Ali et al., 2020; Ari et al., 2016; Dressler & Akan, 2010; Farooq, 2008; Giagkos & Wilson, 2013; Mahdi & Taşpinar, 2023; Rajasekhar et al., 2017; Wagh & Prasad, 2013; Z. Zhang et al., 2013)

<u>Problem Definition:</u>

Current communication networks face challenges related to inefficiency, lack of adaptability, and suboptimal coordination. The need for improved communication systems arises from these shortcomings, where existing technologies may struggle to efficiently transmit information, adapt to dynamic conditions, or maintain robust coordination. This problem statement prompts exploration for innovative solutions that can enhance the efficiency and adaptability of communication networks. Inspired by the highly effective communication systems observed in bee colonies, the objective is to address the limitations of current human-engineered networks by drawing on the successful strategies evolved in nature. The problem, therefore, centers on the optimization of communication networks through biomimicry, leveraging the insights gained from studying bee communication for the development of more efficient and resilient systems.

Fig 15.6: Honey bees in their hives(*Royalty-Free Photo: Two Yelloe Bees | PickPik*, n.d.)

The development of communication systems inspired by bees involves a structured approach:

Step 1: Define the Problem

Identify challenges or ine ciencies in existing communication networks. Recognize the need for optimization and improved efficiency in communication systems. Define a problem statement that addresses these challenges and sets the stage for biomimetic exploration inspired by bees.

Step 2: Biological Research

Conduct thorough biological research on the communication systems within bee colonies. Explore how bees efficiently transmit information, coordinate activities, and maintain a robust network. Understand the mechanisms, signals, and organizational structures employed by bees for effective communication.

Step 3: Analogous Inspiration

Draw inspiration from the biological findings to develop analogous communication strategies. Translate the communication mechanisms observed in bee colonies into engineering principles suitable for human-designed communication networks. Identify key features that contribute to the efficiency of bee communication.

Step 4: Abstraction

Abstracting the fundamental principles derived from bee communication, distilling them into essential design concepts. Identify the core elements that enhance efficiency in bee communication systems. This involves simplifying the biological model to extract principles applicable to human-engineered communication networks.

Step 5: Application

Applying the abstracted principles to optimize human-designed communication systems. Implement biomimetic strategies inspired by bee communication, incorporating them into the architecture and protocols of future communication networks. The goal is to create more efficient, adaptive, and resilient communication systems by emulating the success of nature's bee colonies.

EXAMPLE 5. Self-healing Materials Inspired by Skin: Researchers may develop materials with self-healing properties, similar to the way skin repairs itself, providing longer-lasting and more resilient products. (Aïssa, 2014, 2014; Banshiwal & Tripathi, 2019; Gopalakrishnan & Mishra, 2023; Haines-Gadd et al., 2021; Kong et al., 2023; Perera & Coppens, n.d.; M. Wang et al., 2021; X. Zhao et al., 2017; Zhong & Post, 2015)

Problem Definition:

Traditional materials often lack the ability to autonomously repair damage, leading to reduced product lifespan and resilience. The challenge lies in creating materials that can self-heal, mimicking the regenerative properties observed in natural systems such as skin. The problem is defined by the need for longer-lasting and more resilient products that can autonomously repair damage over time. Inspired by the self-repair mechanisms inherent in skin, the objective is to develop innovative materials capable of self-healing, thereby addressing the limitations of conventional materials in maintaining durability and extending the lifespan of products.

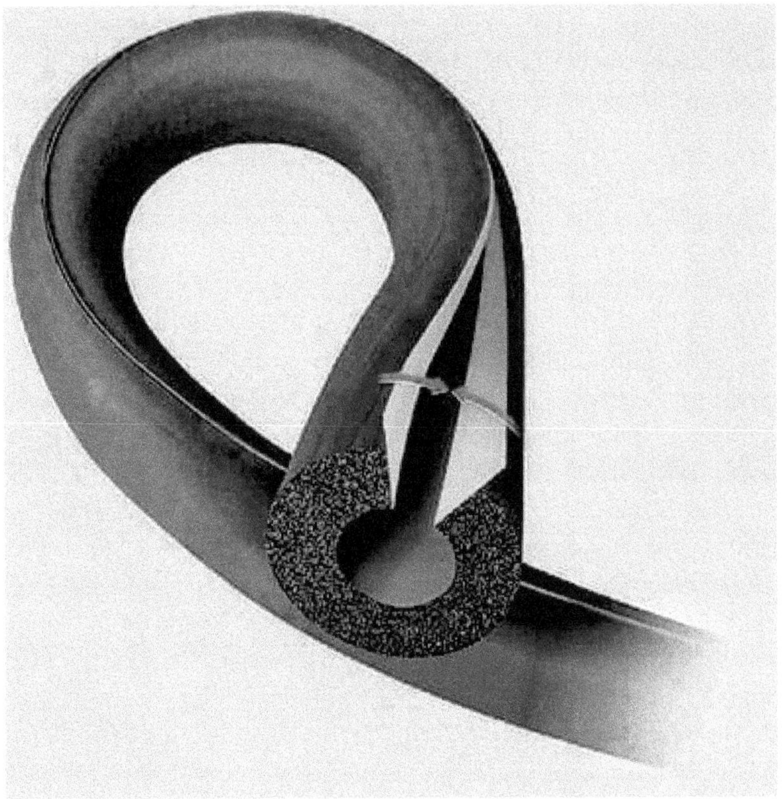

Fig15.7: An image depicting self healing nature of a pipe(*Top-Rated External Pipe Insulation - At Cheapest Prices*, n.d.)

Step 1: Define the Problem

Recognize the limitation of traditional materials in terms of longevity and resilience due to their inability to self-heal. Define the need for materials that can autonomously repair damage, providing longer-lasting and more resilient products.

Step 2: Biological Research

Conducting a thorough biological research on the self-healing properties of skin. Explore how skin naturally repairs itself through regenerative processes, understanding the molecular and structural mechanisms involved in the healing of wounds and damage.

Step 3: Analogous Inspiration

Drawing inspiration from the biological findings to develop analogous solutions for self-healing materials. Translate the regenerative principles observed in skin into engineering and design concepts. Identify key features and mechanisms that contribute to the self-healing properties of skin.

Step 4: Abstraction

Abstracting fundamental principles from skin's self-healing mechanisms. Distill these principles into essential design concepts for creating self-healing materials. Identify the core elements that enable autonomous repair and resilience in natural systems.

Step 5: Application

Applying abstract principles to the development of self-healing materials. Implement biomimetic strategies inspired by skin's self-repair mechanisms, creating materials that can autonomously repair damage over time. The goal is to provide practical solutions for longer-lasting and more resilient products, addressing the defined problem through biomimicry.

EXAMPLE 6. Efficient Water Collection Inspired by Desert Beetles: Designing surfaces that mimic the water-harvesting abilities of the Namib Desert beetle could lead to more efficient water collection methods in arid environments. (Aslan et al., 2022; Bhushan, 2019, 2020; Bu & Bai, 2023; Gurera & Bhushan, 2019, 2020; He et al., 2023; Q. Wang et al., 2021; S. Zhang et al., 2017)

The problem addressed by designing surfaces inspired by the Namib Desert beetle's water-harvesting abilities is as follows:

Problem Definition:

Arid environments, characterized by low water availability, pose significant challenges for efficient water collection. Conventional methods may be inadequate in harvesting water from the atmosphere in these harsh conditions. The need arises for innovative solutions that can improve water collection efficiency in arid regions. Inspired by the Namib Desert beetle, which has evolved unique adaptations for harvesting water from fog-laden air in the desert, the problem statement focuses on designing surfaces that replicate these efficient water-harvesting abilities. The objective is to create biomimetic surfaces capable of collecting water from the atmosphere, providing a sustainable and effective solution for water scarcity in arid environments.

While these examples are speculative, they showcase the potential for biomimicry to drive innovation in various fields in the future.

Fig.15.8: A desert beetle(Kennedy, 2012)

<u>Step 1: Define the Problem</u>

Recognizing the challenge of in efficient water collection in arid environments, where conventional methods fall short. Define the need for a biomimetic solution to improve water harvesting, inspired by the efficient mechanisms observed in the Namib Desert beetle.

<u>Step 2: Biological Research</u>

Conducting an in-depth biological research on the water-harvesting abilities of the Namib Desert beetle. Understand the beetle's adaptations, such as specialized surfaces and behaviors, that enable it to efficiently collect water from fog-laden air in the desert environment.

<u>Step 3: Analogous Inspiration</u>

Drawing inspiration from the biological findings to devise analogous solutions for designing surfaces. Translate the unique adaptations of the Namib Desert beetle into engineering principles, seeking to replicate the water-harvesting abilities of the beetle in artificial surfaces for more efficient water collection.

<u>Step 4: Abstraction</u>

Abstracting the fundamental principles from the Namib Desert beetle's water-harvesting abilities. Identify key features and design concepts that contribute to effective water collection. Distill these elements into essential guidelines for creating biomimetic surfaces capable of emulating the beetle's efficiency.

<u>Step 5: Application</u>

Applying the abstracted principles to the design and development of surfaces for water collection. Implement biomimetic strategies inspired by the Namib Desert beetle, creating artificial surfaces that mimic its efficient water-harvesting mechanisms. The goal is to provide a practical solution for enhancing water collection in arid environments, addressing the defined problem through biomimicry.

Summary

This chapter outlines the methodology of biomimicry, which involves using nature's solutions to address human challenges. The process consists of the following steps:

1. Define the Problem: Clearly identify the issue to be addressed.

2. Biological Research: Study relevant organisms and systems in nature to understand how they solve similar problems.

3. Analogous Inspiration: Identify natural models that offer solutions applicable to the human problem.

4. Abstraction: Extract key principles from the natural models.

5. Application: Implement these principles into human designs and technologies.

6. Testing and Refinement*: Test the biomimetic designs, gather feedback, and make improvements.

7. Iterate: Repeat the process to refine the design through multiple iterations.

The chapter also discusses speculative future applications of biomimicry, including:

1. Energy Harvesting Inspired by Plants: Enhancing solar energy technologies by mimicking photosynthesis.

2. Adaptive Camouflage Inspired by Octopuses: Creating materials that can change color and texture for better camouflage.

3. Structural Materials Inspired by Spider Silk: Developing stronger and more flexible materials by replicating the properties of spider silk.

4. Communication Systems Inspired by Bees: Improving communication networks by adopting strategies from bee colonies.

5. Self-healing Materials Inspired by Skin: Designing materials that can autonomously repair themselves.

6. Efficient Water Collection Inspired by Desert Beetles: Developing surfaces that can collect water efficiently in arid environments, inspired by the Namib Desert beetle.

References

Aghakhani, S., Larijani, A., Sadeghi, F., Martín, D., & Shahrakht, A. A. (2023). A Novel hybrid artificial bee colony-based deep convolutional neural network to improve the detection performance of backscatter communication systems. *Electronics, 12*(10), 2263. https://www.mdpi.com/2079-9292/12/10/2263

Aïssa, B. (2014). *Self-healing Materials: Innovative Materials for Terrestrial and Space Applications.* Smithers Rapra.

Ali, A., Hafeez, Y., Hussainn, S. M., & Nazir, M. U. (2020). Bio-inspired communication: A review on solution of complex problems for highly configurable systems. *2020 3ʳᵈ International Conference on Computing, Mathematics and Engineering Technologies (iCoMET)*, 1–6. https://ieeexplore.ieee.org/abstract/document/9074143/

Araque, K., Palacios, P., Mora, D., & Austin, M. C. (2021). *Biomimicry-Based Strategies for Urban Heat Island Mitigation: A Numerical Case Study under Tropical Climate.*

Glossary Of Terms

1. **Biomimicry:** The practice of imitating nature's solutions to solve human problems.
2. **Biomimetic Design:** Design inspired by nature's principles and patterns.
3. **Bioinspiration:** Drawing inspiration from nature to innovate and solve human challenges.
4. **Adaptation:** A trait or behavior that helps an organism survive and reproduce in its environment.
5. **Evolution:** The process by which species change over time through genetic variation and natural selection.
6. **Ecosystem:** A community of organisms interacting with each other and their physical environment.
7. **Biodiversity:** The variety of life forms in a particular habitat or ecosystem.
8. **Sustainability:** Meeting the needs of the present without compromising the ability of future generations to meet their own needs.
9. **Functional Morphology:** The study of how the form and structure of organisms relate to their function.
10. **Biomaterials:** Materials derived from living organisms or inspired by biological systems.
11. **Biomechanics:** The study of the mechanical aspects of biological systems, such as movement and forces.
12. **Ecological Footprint:** The impact of human activities on the environment, measured in terms of resources consumed and waste generated.
13. **Mutualism:** A symbiotic relationship between two species in which both benefit.
14. **Mimicry:** The resemblance of one organism to another or to its surroundings for protection or other advantages.
15. **Biophilic Design:** Design that incorporates elements of nature to create a more harmonious and productive environment for humans.
16. **Analogous Structures:** Structures in different species that have the same function but evolved independently, often due to similar environmental pressures.

17. **Convergent Evolution:** The independent evolution of similar traits in unrelated species due to similar environmental pressures.
18. **Divergent Evolution:** The accumulation of differences between groups of organisms, leading to the formation of new species.
19. **Homologous Structures:** Structures in different species that are similar because they are inherited from a common ancestor.
20. **Ecomimicry:** The practice of mimicking entire ecosystems or ecological processes to address human challenges.
21. **Biomimetic Materials:** Materials that mimic the properties or structures found in nature, such as self-healing polymers or superhydrophobic surfaces.
22. **Bioremediation:** The use of living organisms to remove or neutralize pollutants from a contaminated environment.
23. **Nanobiotechnology:** The intersection of nanotechnology and biology, often involving the manipulation of biological molecules or systems at the nanoscale.
24. **Genetic Engineering:** The manipulation of an organism's genetic material to produce desired traits or outcomes.
25. **Bio-inspired Robotics:** Robotics that draw inspiration from biological systems to improve performance and efficiency.
26. **Biomineralization:** The process by which organisms produce minerals, often resulting in complex structures like shells or skeletons.
27. **Bioinformatics:** The use of computational tools and techniques to analyze and interpret biological data, such as DNA sequences.
28. **Biofeedback:** The use of biological signals, such as heart rate or brain activity, to provide feedback for self-regulation and control.
29. **Biomimetic Synthesis:** The production of materials or compounds using methods inspired by biological processes, such as photosynthesis or enzyme catalysis.
30. **Ecological Engineering:** The design of sustainable systems that integrate human society with natural ecosystems to benefit both.

About The Authors

- **Mr. Yagesh Kumar:**

The author, Yagesh Kumar, holds a Dual Degree (BTech-MTech) in Mechanical Engineering from the Indian Institute of Technology, Ropar. With experience in both academia and industry, Yagesh has engaged in projects ranging from biomimicry to electric vehicle analysis. His expertise includes programming languages such as C and Matlab, along with software proficiency in ABAQUS and Solidworks. Key courses taken encompass various aspects of mechanical engineering, from modern manufacturing to analysis of material processes.

- **Dr. Prabir Sarkar:**

Dr. Prabir Sarkar is an Associate Professor in the Department of Mechanical Engineering at the Indian Institute of Technology Ropar. He is the founder chairman of the Intellectual Property Rights (IPR) cell of IIT Ropar. Dr. Sarkar's research interests include product design, sustainability, creativity, biomimicry, and design research. He has authored more than 80 peer-reviewed journal and conference publications. His research group in the Design Research Laboratory and Sustainable Design and Manufacturing Laboratory are engaged in research in Ecodesign, biomimicry, engineering aesthetics, sustainable machining, and design creativity. Recently he is also working on developing products for rural applications.